SCIENCE AND ENGINEERING P

The Scientific Community

SIR JAMES TAYLOR

OXFORD UNIVERSITY PRESS 1973

Oxford University Press *Ely House, London, W.I*

Glasgow New York Toronto Melbourne Wellington Cape Town Ibadan Nairobi Dar es Salaam Lusaka Addis Ababa Delhi Bombay Calcutta Madras Karachi Lahore Dacca Kuala Lumpur Singapore Hong Kong Tokyo

Printed in Great Britain by
J. W. Arrowsmith Ltd., Bristol, England

Preface

The past hundred years have witnessed the emergence and spectacular growth in numbers of professional engineers, scientists, and technologists, to become a major social and economic factor in our society. The increase in their numbers during this period, starting from a very low base, has been exponential. Such a growth rate cannot be maintained indefinitely, and it looks as if the levelling out of demand may now be upon us and is partly responsible for some of our present difficulties.

The twentieth century has witnessed a growth, not only in numbers, but in the nature of the individuals comprising the scientific community. Initially, they were mainly self-trained, or of part-time scientific and technical education, brought up in industry and factories. Few of them had a university education. There was consequently no conflict with their environment. Nowadays, practically all scientists, and a high proportion of technologists and engineers, are university graduates. Not surprisingly this has induced a number of environmental conflicts, for the aims and objects of universities and industries are poles apart.

The past hundred years has seen further great changes. Science has emerged as a profession with its own standards, motivations, and attitudes, quite different from those of technology and engineering. Nevertheless, scientists have been drafted into technology in large numbers, particularly during the two world wars. Research and development has become a huge, continuous professional occupation, which is uniquely characteristic of the recent years of the present century. Pure science research has become increasingly dependent on technology which, in its turn, is increasingly dependent on engineering. The universities, which only recently adapted themselves to science, are now faced with the more difficult task of adapting to technology and engineering. The resulting conflicts have not so far been resolved.

On top of all this, western societies, having achieved relative material abundance, have become aware of the price they have paid for this abundance. To quote Smith, R.I.C. External Relations Officer (Smith 1971).

> Environmentalists and sociologists wax eloquent on the dangers and desolation wrought on the social and material environment by science and technology . . . In the middle of all this stands the scientist, embattled, confused, desirous of rethinking his place in current society, worried if there will indeed be a place for him in the future.

People, especially the young, are disenchanted with science and technology, and are turning away from it. Thus, the community is deprived of many of the best brains which could sustain and elevate it. There is severe fragmentation and uncertainty within the scientific and technical community, which is in desperate need of a new ethos which will unite it in meaningful aspiration and purpose. The origin of these conflicts goes back to the distant past, but the problems have only recently become significant, owing to the intensive application of science and technology and the vast increase in qualified engineers, scientists and technologists, during the present century.

The intention of this book is to examine how the scientific and technological community has grown and organized itself into learned societies and how it has aspired to and acquired professional status. I also examine its motivations and functions and the origins of its conflicts and try to point the way to the new horizons which lie before it.

Much of the material is derived from lectures and addresses I have given to the British Association, the Royal Swedish Academy of Engineering Sciences, the Institute of Physics and the Physical Society, the Royal Society of Arts, and the Newcomen Society in North America; from work in connection with the merger of the Institute of Physics and the Physical Society and the recent merger of the Chemical Society, the Royal Institute of Chemistry, the Faraday Society, and the Society for Analytical Chemistry. I am most grateful to these bodies for the many opportunities they have afforded me to gain experience in the various fields with which this book is concerned.

J. T.

Seale, Surrey
February 1972

Contents

1

The ancient heritage

SCIENCE — or its ancient equivalent — was probably first pursued for magical purposes, particularly by priests concerned with religious rites that depended on repetitive positions of the heavenly bodies. This gave rise to astrology, the forerunner of astronomy. Science was for a long period a matter of patronage; in alchemy to find the philosopher's stone, or transmutation of base metals to gold; in military engineering to advance the arts of war; in seafaring to improve navigation; and in medicine to advance the practical art of healing. In addition, there has always been a small number of gifted amateurs motivated by intellectual curiosity.

The Greeks in particular recognized and established science of the day as a cultural pursuit. So did Cicero, who in enumerating the various heads of mental excellence, remarked, 'This pertains most of all to human nature for we are all drawn to the pursuit of knowledge; in which to excel we consider excellent.'

There is a common belief that the early Greeks were entirely theoretical in their methods. The Countess of Albemarle, in an address to the British Association, remarked on this aspect of Greek philosophy. 'With the Greeks science was a *logos* — a philosophy apart from the things of everyday life and outside the knowledge and affairs of ordinary people — it played no part in the political scene.'

This view does less than justice to the Greek contribution to science. Kitto (1969) in *The Greeks* has a section on the Greek mind which puts the Greek contribution in proper balance. The Greeks had a firm belief in reason. 'It is as if the human mind for the first time took its toes off the bottom and began to swim, and to swim with astonishing confidence.' The fact that this vigorous new doctrine flourished in Greece does not mean that the

Greeks were not also practical men. To give but one example from Hodges' *Technology in the Ancient World:*

> The Greeks raised the level of technology from a squatting to sitting position, both metaphorically and physically speaking. It is not clear why, but it did mean that certain crafts — such as woodworking, masonry and fine metal-working — which are carried out better on a bench than at ground level, could evolve further. This in its turn meant the development of new iron tools.

The use of *logos,* according, allegedly, to Hippocrates of Cos, 'is all very well with impenetrable mysteries like what goes on in the sky, or under the earth, but it is not the way with a "craft" (or an "art")' (Kitto 1969). Kitto also quotes from the *Precepts* of Hippocrates.

> In medicine one must pay attention not to plausible theorizing ('logismos'), but to experience and reason ('logos') together . . . I agree that theorizing is to be approved, provided that it is based on facts, and systematically makes its deductions from what is observed . . . But conclusions drawn by the unaided reason can hardly be serviceable; only those drawn from observed fact.

The error of the Greeks was in not realizing that the methods of observation and deduction were of universal application and should not be restricted to arts and crafts. In fairness it must be observed that many of the phenomena of nature which preoccupied the Greek speculations were not open to such methods because no apparatus, such as the telescope, was available to make observations.

In respect of the assumed superiority of the Greek civilization compared with others, Kitto remarks that 'It would be one answer, a true and sufficient one, to say that the older civilizations of the East were often extremely efficient in practical matters and, sometimes, in their art not inferior to the Greeks, yet they were intellectually barren' (Kitto 1969, p. 8). The assumption is that other civilizations made empirical advances in the practical arts and crafts but did not back their observations by systematic logical speculation. Speculation, such as it was, was related to magic, or religion, and scientific progress is at best very slow in such circumstances. Moreover, in the ancient world generally, cultures were respected by the governing classes who were their patrons, but crafts, which were the technics of the day, were not held in any esteem.

A classical example of this attitude occurs in the Apocrypha. *Ecclesiasticus,* chapter 38, dealing with the crafts and skills of the ploughman, carpenter, engraver, smith, and potter, states: 'All these trust to their hands: and everyone is wise in his work. Without these cannot a city be inhabited . . .'; and later in the same chapter: 'But they will maintain the state of the world, and their desire is in the work of their craft.'

Notwithstanding the recognition of the necessity for technics, no status was accorded to the practitioners, for the author of Ecclesiasticus wrote: 'They shall not dwell where they will, nor go up and down. They shall not be sought for in public counsel, nor sit high in the congregation; they shall not sit on the judges' seat, nor understand the sentence of judgment'.

Astronomy, approached through astrology, was probably the first systematized science, but chemistry would appear, in its empirical form, to have been the earliest physical science which received systematic study. Chemistry grew out of alchemistry (alchemy) which originally probably related to the art of extracting juices from plants to prepare potions and medicines, but later became restricted to denote the art of transmutation of metals and the quest for the 'Philosopher's stone'.

The concerns of ancient chemistry, such as Greek fire and gunpowder, transmutation, and the Philosopher's Stone, ensured that chemists had wealthy and influential patrons, notwithstanding opposition by the Church. The investigation of another side of chemistry, namely that related to medicine, also received continued interest from the medical fraternity — one of the few educated classes, even if riddled with quacks.

The motivations of the Ancients were obviously diverse, ranging from the pursuit of knowledge for its own sake (a concept of the Greeks) to the utilitarian motivation of the Arabs, Egyptians, and Romans.

In actual fact there never have been, and never will be, clean-cut differentiations between arts, crafts, technology, and science. To go from the general to the particular, Leonardo da Vinci was an artist, engineer, inventor, and technologist, as well as a scientist. Like many of his contemporaries, Leonardo was concerned with military engineering and, although this activity was probably that which appealed most to his patron, the Duke of Milan, and gained him his living, he was proud of his status as an engineer (Mumford 1967).

The fragmentation of human observation and manipulation of nature into watertight compartments arises from man's chronic habit of labelling everything and erecting organization edifices based on status, rather than function.

2

The new learning and the Royal Society

The Renaissance in Europe was based on the retrieval of Greek and Roman classics, and this had a profound influence on the future of science. Myers (1969) comments:

> Owing largely to the way in which Greek philosophy was recovered, beginning with the logical commentaries of Boethius in the ninth century, or earlier, the West attached great importance to dialectic as a means to the discovery of truth; and the aristocratic complexion of society predisposed it to accept the Greeks' trust in abstract reasoning, rather than the manual labour of experiment, as a path to wisdom.

The Middle Ages in Europe saw the development of universities as bodies with the responsibility for learning. Students and teachers alike, with few exceptions, were clerks in holy orders, and the universities were dominated by the Church. The Bishop of Durham, Dr. I. T. Ramsey (1971), has pointed out that ecclesiastical dogma at that time taught 'that theology is prescriptive, able to determine *a priori* culture and society'. Science could advance only where it did not clash with religion and this placed it in a strait-jacket in which it was held for centuries. Furthermore, the separation of science and religion which was enforced from medieval times led to science and morality being divorced as well.

The prescriptive approach to knowledge was undermined, but not displaced, by the experimental philosophy pioneered by Copernicus, Kepler, Francis Bacon, Gilbert, and Galileo in the Elizabethan era and the seventeenth century. The new philosophy combined the logical methods of the Greeks with accurate practical observations and experiments without preconceived notions as to explanation. It was a characteristic of the era of new learning that it and religion were kept in watertight compartments; indeed, many of the new scientists (Newton for example) were deeply religious.

The new learning certainly owed much to enlightened university graduates, but not to the universities as such, although mathematics was taught at universities and provided it with tools. One of the first scientific subjects to be taught was chemistry, so far as it was required in medicine. The University of Aberdeen, for example, had its first chair in chemistry, it is claimed, in 1505; it was termed 'Medicines'. The present chair was founded in 1793. Cambridge, Dublin, and Edinburgh set up chairs of chemistry in 1702, 1710, and 1713 respectively, but most of the university chairs were much later.

The dependence of science on technics in the seventeenth century is well illustrated by Mumford (1967), who observed that the invention of the microscope and the telescope in the seventeenth century rested upon the improvement of glass-manufacture and that 'these inventions opened up a new world both of micro-organisms and of distant stars and galaxies'. He adds: 'The increase of the scope of scientific discovery and the tempo of mechanical invention can both be largely attributed to the printed book, and from the seventeenth century on, to the printed scientific paper and review'. The third invention which transformed scientific method was the clock, which dates from the fourteenth century, but was perfected in the sixteenth century.

One of the results of the Renaissance which re-established man's interest in the humanities and natural knowledge was to bring men of like interests together for discussion, first in informal groups and later into more formal associations, the learned societies. To promote these interests, academies of learning were formed, particularly in Italy, where the movement started. Examples are the Accademia dei Lincei and the Accademia del Cimento, which functioned between 1600 and 1660. These academies derived their names from the grove near Athens where Plato (427–347 BC) and his followers held their philosophical discussions.

Early attempts to found academies in England were unsuccessful, and it was not until 1660 that a really influential and permanent society was founded, The Royal Society.

The *Record of the Royal Society of London* (Royal Society 1912) claims that the Society was one of the earliest practical fruits of the philosophical labours of Francis Bacon. 'His aim was to enforce the patient investigation of nature by observation and experiment.' In 1617 in his *New Atlantis* he proposed an Academy, or College, with 36 Fellows to carry out research. Bacon died in 1626 and the *New Atlantis,* which was first published a year after his death, went through ten editions in 43 years, which shows how seriously the new learning was taken by his contemporaries.

Any group of people can come together informally to form a society and draw up regulations for the conduct of their mutual affairs, and many such societies existed before the Royal Society. Such informal bodies suffer certain disadvantages so far as protection by law is concerned (see pp. 21–2) and as soon as they regard themselves as reasonably established they usually seek incorporation. Incorporation in this context means to constitute into a corporation recognized by law, with specific functions, duties, rights, and liabilities.

The Royal Society itself was the final achievement of groups which had met informally for a considerable number of years. In about 1645 a number of men of similar ideals, among whom were Robert Boyle, John Wilkins, John Wallis, John Evelyn, Robert Hawke, Christopher Wren, and William Petty, met together to carry out experiments and discourse. This 'invisible college' met in private houses and taverns and its deliberations excluded theological and state affairs. These and similar meetings led to the formation of the Oxford Society which met until 1690. Its counterpart, the London Society, held meetings until about 1658, usually at Gresham College.

In February 1660 Monk and his army entered London, and Charles II mounted the throne at the end of May the same year. A period of relative stability ensued and meetings of the philosophers were resumed. The advisability of a more formal organization by incorporation was soon realized and on 18 November 1660 a memorandum for a constitution was drawn up for the founding of a College for the promoting of Physics — Mathematical Experimental Learning. A list of 41 names was prepared and recorded for the membership of the proposed new body and a petition for a royal charter was submitted to the King.

Incorporation in England prior to 1862 was by charter. 'A written document delivered by the Sovereign or legislature: a. granting privileges or recognizing rights; b. creating a borough, university, company, or other corporation.' (*S.O.E.D.*)

The best-known charter is, of course, Magna Carta, the 'Great Charter' granted in 1215 by King John which secured important liberties and rights to every citizen of the Kingdom. Charters were a feature of English medieval life and were much prized. The London Craft Guilds are mainly bodies with ancient royal charters, and the Tudor sovereigns used charters a great deal to confer monopoly trading on companies of adventurers. They have now become a symbol of status and the hallmark of respectability. Charters are the prerogative of the Sovereign who grants them, provided that the Privy Council approves a petition submitted to the Sovereign.

The petition for a charter submitted by the Royal Society was favourably received by Charles II, who, though no scientist himself, was aware of the advantages of patronizing experimental philosophy, and a royal charter in Latin was granted in 1662. The conferring of a royal charter does not carry with it the right to use the word 'Royal', but this privilege was granted to the new Society by the King.

The Royal Society, whose full title is 'The Royal Society of London for Improving Natural Knowledge', is the paramount learned society in Britain. The Fellowship is restricted in numbers. The Journal-Book of the Society records that at a meeting on 12 December 1662 it was decided 'that the stated numbers of this Society be five and fifty' and 'that any person of the degree of Baron, or above, may be admitted as supernumerary . . .' (Royal Society 1912). At the end of 1971 there were 735 British Fellows, 7 Patron and Royal, and 70 Foreign Members.

The Royal Society still functions under the two charters granted by Charles II, the first in July 1662, two years after its foundation, and the second in April 1663. Its methods and procedures are laid down in the 'Statutes of the Royal Society' and the 'Standing Orders of Council'. It has a Council of 21, out of which are chosen the President, Treasurer, two Secretaries, and a Foreign Secretary. Its procedures, methods of election, and activities are fully described in the *Year Book of the Royal Society*, which is issued annually.

The era of organized science in this country had its beginnings in the Royal Society. The charters included within their scope all branches of science, pure and applied, and enjoined its members 'through experiment to deal with the sciences of natural things and useful arts'. The concentration on pure science came much later.

For more than fifty years the Royal Society had the field to itself in science and technology, and for over a century in pure science. The Society of Antiquaries, which was formed in 1717, barely impinged on the Royal Society's activities, and the Dublin Philosophical Society, founded in 1683, on the model of the Royal Society, did not last long.

According to Millerson (1964) both the Royal Society and the Society of Antiquaries 'lost their early enthusiasm during the eighteenth century and declined into uninspiring Gentlemen's Clubs'. Ups and downs in bodies based largely on the enthusiasm of dedicated individuals are, of course, to be expected.

The ferment which finally produced the industrial revolution began to work in the first half of the eighteenth century, accompanied by a growth of trade, commerce, and manufactures, and men began to turn their minds towards organizations of individuals which could promote these activities. The oldest is the Royal Society of Dublin, formed in 1731 with the full title of 'The Dublin Society for improving Husbandry, Manufactures, and other Useful Arts'. This Society was incorporated in 1749 with the title of 'The Royal Dublin Society'.

In America, Benjamin Franklin published (1743) a 'Proposal for promoting useful knowledge among the British Plantations', and this led to the formation of the American Philosophical Society (1790), which still flourishes.

In England the Society of Arts was formed in 1754, its full title being 'The Society for the Encouragement of Arts, Manufactures and Commerce'. Benjamin Franklin was one of its first active members, but the credit for the proposal which founded the (as it is now called) Royal Society of Arts (R.S.A.) rests with William Shipley, a Northampton drawing-master who in 1753 put forward 'proposals for raising by subscription a fund to be distributed in premiums for the promoting of improvements in the liberal arts and sciences, manufactures, etc.' (Trueman Wood 1913). It is said that he got the idea from horse-racing meetings when he had observed that the presentation of plates and prizes stimulated entry and success in horse-breeding. The Society was founded in a London coffee-house by a number of 'Noblemen, Clergy, Gentlemen and Merchants' determined to use their diverse experience and knowledge 'for the Encouragement of Arts, Manufactures and Commerce'. Nowhere else in Europe at the time would you have found 'Noblemen, Clergy, Gentlemen and Merchants' hob-nobbing together: this was a special English feature.

3
The industrial revolution and Victorian England

THE industrial revolution was not the deliberate outcome of an application of scientific knowledge, but of technics in an *ad hoc*, and often serendipital, fashion. There was no scientific gap between England and Europe, but England had that combination of circumstances which gave her the impetus and will to innovate and use what empirical knowledge and technology were available.

The Society of Arts took an impressive initiative, not only in the early stages of the industrial revolution, but in its later, more technological phases. Its strength has always resided in being a generalist society with members from all walks of life and disciplines. Membership was not restricted, since it was only necessary to be in sympathy with the objects of the Society, and membership included females.

The list of the earliest members is very impressive. It embraced not only peers and wealthy men, but (to quote) 'Statesmen, philosophers, philanthropists, painters, lawyers, divines, physicians, authors, dramatists, actors, musicians, bankers, soldiers, sailors, architects, historians, mechanicians, merchants, all are to be found, and many of them the most eminent of the time'.

The first printed list of members was issued in October 1755 and comprises 110 names. The next list (1758) contains 708 names, and that of the following year, 1760. Membership became very fashionable and within a few years it reached 2500. After various vicissitudes the Fellowship now (1972) numbers 9500. The current qualifications for Fellowship (though not written in the Bye-Laws) are a university degree, or its equivalent in achievement.

The R.S.A. is remarkable in that it functioned for nearly a century as an informal non-incorporated association operating on the 'plan' put forward by its first Secretary, William Shipley (Allan

1968). It was not until 1847 that the Society petitioned for, and received, a Royal Charter.

The next great learned society to be formed was the Linnean Society of London, which commemorated Carolus Linnaeus, the Swedish biologist who devised the system of scientific nomenclature for plants and animals now in general use. The Society was founded in 1788 and was incorporated by royal charter in 1802. The aims of the Society, as stated in its Charter, are 'the cultivation of the Science of Natural History in all its branches'. It houses Linnaeus's collection of plants and animals and his library. Sir Joseph Banks was one of its early Honorary Members and Charles Darwin, Joseph Hooker, and Thomas Huxley were all active Fellows. Indeed, Charles Darwin and Alfred R. Wallace had their first papers on evolution read to the Society in 1858.

Another body which has contributed greatly to the advancement of science and the scientific community, the Royal Institution of Great Britain (R.I.), was set up in 1799 and received its first charter from George III in 1800. Its objects were to form 'A Public Institution for diffusing the knowledge, and facilitating the general introduction of useful mechanical inventions and improvements; and for teaching by courses of Philosophic Lectures and experiments the application of science to the common purpose of life'. It has the distinction of being the first scientific research centre in Britain.

Its founders, who were called proprietors and managers, included four earls, which shows the profound interest taken by the educated nobility in science at the beginning of the industrial revolution. Among the managers were also Sir Joseph Banks and Count Rumford. The original Foundation is an unusual one and its terms were modified by Act of Parliament in 1810, which recognized that it had become 'A Public National and Permanent Establishment, devoted to the Cultivation of Practical Science, and to the Promotion of every Improvement in Agriculture, Manufactures, and the Arts'. Its early Directors were Davy and Faraday.

Faraday, born in 1791, came of poor parents of northern stock. Although he was a man of genius, he was of abounding humility, belonging as he did to a small strict set, known as Sandemanians. For him there was no conflict or contradiction between religion and science; they were things apart. In fact, he said, '. . . in my intercourse with my fellow creatures, that which is religious and that which is philosophical have ever been two distinct things' (Riley 1954, p. 3).

The names of Davy and Faraday are commemorated in the Davy Faraday Research Laboratory of the R.I., endowed by Ludwig Mond in 1896. The R.I. is internationally famous, not only for its research, but for its Friday evening discourses and Christmas lectures to children.

The growing public appetite for science in the nineteenth century was catered for in an original and imaginative fashion by the formation of the British Association for the Advancement of Science (B.A.) in 1831, by a group of Fellows of the Royal Society. It remained a voluntary association until 1928 when it was incorporated by royal charter. This period as a non-incorporated body is almost as long as that of the R.S.A., and they must be records. In the early days the B.A. was used as a forum for announcing original discoveries and for discussing recent developments, but this role has been taken over by the specialist societies.

The B.A. is a generalist society, particularly concerned with 'presenting to the public generally and to industry and Government specifically, the ever-increasing importance of science, and its applications, to the country's economic and social development'. The B.A. Annual Meeting has always been its special feature. It is held in a different city each year and lasts a week, covering a wide range of sciences and semi-social activities. To a large number of academics and teachers, and private persons interested in science, it is the highlight of the year.

The early societies, although generalist, were aware of the advantages of specialist knowledge. The R.S.A., for example, had six principal committees which dealt respectively with Agriculture, Chemistry, Polite Arts, Manufactures, Mechanics, and Colonies and Trade. It had a number of distinguished chemists among its members and the Chemistry Committee was one of the most important in the Society. Indeed, it was through the efforts of the R.S.A. that the Chemical Society was founded in 1841.

Great advances in chemical science were made by Lavoisier and Dalton towards the end of the eighteenth century, and by Faraday in both chemistry and experimental physics at the beginning of the nineteenth century which heralded the great age of classical physics. These established the science of chemistry on a firm basis and promoted the formation of what is probably the first specialist or single-subject society. The reasons for this are well described in Warington's description, in 1891, of the foundation of the Chemical Society fifty years earlier (1896, p. 115), as follows:

> It is obvious that a public association for the study and furtherance of any particular science will be formed only when those who study and those who apply that science have become sufficiently numerous to form a considerable body, and when the science itself has developed to such a degree that the field of investigation has become sufficiently wide to accommodate many workers, and sufficiently fruitful to encourage them to attempt greater and more united efforts. The early scientific societies never confined themselves to one branch of science. Our own Royal Society is an example of the wide range of subjects which it was necessary to include during the childhood of science in order to supply enough material to sustain the interest of the Society's meetings.

Even so, there was an objection to the proposal on the grounds that chemistry itself was not enough to support a society and that it ought to be joined to electricity.

The number of Fellows in the Chemical Society, initially (1841) 77, had risen to 551 by 1870. Chemists were, and still are, greater in numbers than other scientists because there has been for well over a century an important and growing chemical industry which cannot operate without them. The Chemical Society was incorporated by royal charter in 1848, its objects being 'for the general advancement of chemical science, as intimately connected with the prosperity of the manufactures of the United Kingdom'. Its membership was exclusively male.

A considerable number of societies arose in the 1840s and 50s, indicating the growth of an educated public with a wide interest in natural science. Examples are the Philological Society (1842); the British Archaeological Association (1843); the Royal Meteorological Society (1850); the Royal Photographic Society (1854); the Geologists' Association (1855); and the British Ornithologists Union (1858). This was the age of Samuel Smiles's 'Self Help' when what the Americans call self-enhancement was at its peak. It was the age of mechanics' institutes, literary and philosophical societies, and night-classes.

The importance of learned societies in fulfilling objects in the public interest was recognized by Government from the early 1840s when they were accorded privileges, as in the case of the

Royal Society, the Royal Academy, the Chemical Society, and the Linnean Society, of grace-and-favour accommodation — currently at Carlton Terrace and Burlington House. In addition, most learned societies rank in law as charities and enjoy privileges in respect of taxation and rates. To-day, if all disciplines (scientific and non-scientific) are included, there are at least 800 bodies which qualify as scientific and learned societies in Great Britain (Anon. 1964).

4

Professional aspirations and the qualifying associations

THE industrial revolution was accompanied by a rise in the numbers and importance of the English middle classes and a growing pressure by them for status. This resulted in the professional classes, as we know them, and the professionalization of engineers and scientists.

Until the middle of the last century positions which would now be regarded as professional occupations were filled mainly, if not entirely, by people who obtained them by some form of privilege or patronage, not related to qualifications. Certain offices of Church, or State, were in the gift of patrons; others were obtained by influence, or by direct purchase. There were, until last century, if one excepts the Armed Forces, only three recognized professions: the Church, medicine, and the law. Reader (1966) describes the rise of the professional classes in nineteenth-century England. It is a fascinating account of which only a short summary can be given here.

The upper classes did not engage in trade if they could possibly help it, and 'the view of science as a fit hobby for well-bred amateurs, but nothing more, persisted long in England' (Reader 1966). The favoured occupations of the gentry were 'Government and War', with the Church following, and law and medicine a long way down.

The only education which fitted entrants for jobs of authority was a classical one compounded of Greek and Latin, classics, and Euclidean geometry. Any other sort of education was regarded with contempt. Indeed, there was no real professional training, even for the Church, law, or medicine.

The rising middle classes resented the system because they were excluded from it, and a long campaign began to make entry into professions by qualification instead of preferment and purchase. The qualification envisaged was not that which 'depended on eating the right number of dinners and paying the right fees' (Reader 1966, p. 45), but those which depended on competitive examinations, or what have become known as 'paper qualifications'.

The ethics of professionalism developed slowly and arose from the Victorian virtues of 'earning a living; raising the moral tone of society and social climbing' (Reader 1966, p. 1). The real breakthrough occurred in the Civil Service when in 1853 competitive examination for entry into the Indian Civil Service was established. The system gradually extended to departments of the Home Civil Service and, finally, in 1870, the method of entry by open competition was established by Order-in-Council.

The open competition system required new concepts in respect of education and the Indian Civil Service Committee's proposals meant a shift from the traditional training to one which included English and history, various foreign languages, and natural science, as well as classics and mathematics. The object was to provide equal opportunities for candidates from all over the country, as well as from Oxford and Cambridge.

The open competition system for Civil Service appointments was accompanied, or followed by, a rising demand for qualifications on the basis of written examinations for other professions, such as law and medicine, and younger professions, like accountancy. The demand finally had an effect on school numbers and curricula.

The old-established method of training in Britain was by apprenticeship, and in the professions pupils were indentured to established practitioners. The examination qualifications were somehow fitted into the practical work by part-time studies; indeed, Victorian England developed the finest night-school and part-time educational facilities in the world, and this accounts for much of its success. It was a long haul before professional qualifications could be achieved by university education alone.

Well-known engineers and architects attracted many pupils, and so did the great engineering firms. The apprentice system, even for young professions such as chemistry, lasted into the twentieth century. City analysts and consultants, such as the famous Dr. Dunn at Newcastle upon Tyne (Chirnside 1971), were much sought after to train pupils who came up the profession by pupilage under indentures, together with examination by the Institute of Chemistry. Indeed, almost up to the Second World War, chemical laboratory assistants were indentured apprentices in the research laboratories of Nobel's in Scotland, which had its own arrangements for scientific education leading to professional qualifications in the Institute of Chemistry.

Qualification by written professional examinations was a new conception, but professional organizations as such have a long history, the first being the Inns of Court, founded in about 1400. Others of various sorts were formed by barbers, surgeons, and physicians from the late fifteenth century onwards (Millerson 1964, p. 17). Modern-type professional and qualifying associations were, however, the product of the nineteenth century.

Pride of place can be claimed by the Institution of Civil Engineers, formed in 1818, which superseded the earlier Society of Civil Engineers formed in 1771 for the 'closer communication of ideas'. The objects of the Institution were 'facilitating the acquirement of knowledge necessary in the civil engineering profession and for promoting mechanical philosophy'.

At that time there were only two classes of engineers, military and civil. Civil engineers included all types of engineers, including people like Newcomen, Boulton, and Watt, who were concerned with steam engines.

Other qualifying bodies formed in the next few decades were the Pharmaceutical Society, formed in 1841 by chemists and druggists, and the Institution of Mechanical Engineers, founded in 1847.

Professional bodies differ from learned societies in a number of important respects, although many of their activities overlap. Learned societies exist to promote and forward the disciplines which they represent, whereas professional bodies are concerned with promoting the profession and interests of their members, as well as the subjects with which they deal.

Entrance to membership of a learned society is not dependent upon formal examinations, but is by selection. Professional bodies require formal qualifications for membership, and most, but not all, have their own qualifying examinations, corresponding to the various grades of membership.

Membership of some bodies, e.g. the Pharmaceutical Society, the Royal College of Medicine, and the Royal College of Surgeons, conveys the right to practice. This is not the case in the scientific qualifying associations, which are concerned with subjects that require no formal licence for practising the profession.

Apart from chemists, the scientific community hardly existed until the latter part of the nineteenth century. There were a few outstanding scientists whose influence was great, and there were also 'gentlemen of scientific tastes'. Such gifted amateurs made a great contribution to the progress of science, but they were relatively few in numbers. Science had to be institutionalized before it could become a profession (Shils 1968).

Census figures of some reliability go back to 1851. This was the date of the great International Exhibition at the Crystal Palace, when Britain was at its industrial peak. The 1851 Census lists the category of 'Scientific Persons', as do those of 1861 and 1871. The 1881, 1891, 1901, and 1911 Censuses changed the category to 'Persons engaged in Scientfic Pursuits'. Thereafter, the scientific occupations are further categorized. Fortunately, for our present purposes, no great accuracy is required, as we shall be concerned only with broad trends. Using the information available, and excluding certain categories, such as laboratory assistants and technicians, where information is available (as, for example, in the 1951 Census), rough figures for the number of scientists from 1851 onwards can be obtained.

From these figures it will be seen that the number of persons engaged in scientific pursuits in 1851 comprised 418 males and 24 females. In 1861 it had risen to 637 and 42 respectively; in 1871 to 909 and 49.

By 1871 the Fellowship of the Chemical Society had risen to 551 and by 1881 to 1024. This does not mean that half the scientists in 1870 were chemists, since many chemists engaged in manufacture would not be classified in the Census as engaged in scientific pursuits. Nevertheless, since many of the chemists in manufacture would not be members of the Society, it does indicate a very considerable preponderance of chemists.

5

Chemistry becomes a profession—Physics a discipline

As might be expected, because of their relatively great numbers, the chemists were the first scientists to be professionalized and form a qualifying association.

Pilcher (1914), a former Registrar and Secretary, gave an interesting account of the history of the Institute of Chemistry of Great Britain and Ireland from 1877 to 1914.

At a meeting in the rooms of the Chemical Society on 27 April 1876 to discuss the organization of the chemical profession, a committee of ten was appointed to discuss with the Chemical Society the possibility of forming an organization of professional chemists within the Chemical Society, the suggested title being 'The Institute of Professional Chemists of Great Britain and Ireland'. Counsel's opinion was taken in respect of the powers of the Chemical Society under its Royal Charter, and the Chemical Society was advised that its charter would not permit it to add to its name or sanction the creation of an institute as an adjunct to the society. It was decided therefore to form a new association. A subcommittee drew up a 'Draft Scheme' and presented it on 24 February 1877. The aims of the proposed body were the maintenance of the status and efficiency of the profession of chemistry.

The Memorandum and Articles of Association were drawn up and application for incorporation was made to the Board of Trade. Trouble then arose. It was pointed out that in the eye of the law, as it then stood, chemists were not recognized as 'chemists', which in law were pharmaceutical chemists and chemists-and-druggists already represented by the Pharmaceutical Society. In the end the objection of the Board of Trade, which stemmed from advice from the Privy Council and the Pharmaceutical Society, was got over by giving a new name to the proposed body, namely, 'The

Institute of Chemistry of Great Britain and Ireland' and, after various amendments of the Memorandum and Articles, the Institute of Chemistry of Great Britain and Ireland became incorporated on 2 October 1877 by licence by the Board of Trade under the provisions of Section 23 of the Companies Act 1867. The initial membership was 225 (February 1878). The current membership (1971) is 27000.

It will be observed that the Institute of Chemistry was incorporated initially, not by charter, but by licence of the Board of Trade. It presented a petition for, and obtained, a royal charter in 1885 under the title of 'The Institute of Chemistry of Great Britain and Ireland'. A supplemental charter was obtained in 1944 permitting the word 'Royal' to be used before the previous title, and the Institute became the 'Royal Institute of Chemistry of Great Britain and Ireland'. A further royal charter was granted in 1949 introducing a number of amendments which were appropriate to the conditions of the time and the name was changed to 'The Royal Institute of Chemistry' (R.I.C.).

Originally membership was restricted to British (later including ex-British to cover Commonwealth Countries) but this restriction was removed in 1966 when the bye-laws were amended.

To many people the structure and constitutions of learned societies and qualifying associations appear unduly complicated. As mentioned earlier, any group of individuals with common interests can unite together informally to form a society, and many such exist. There are, however, perils and disadvantages in the informal status, which most societies are not prepared or able to accept, and most are incorporated in one fashion or another. According to Millerson (1964), the main disadvantages are that such an association cannot sue, or be sued, in its own name; property can be held only on 'deed of trust' by appointed trustees; officers of the body are not legally protected and may be proceeded against on a basis of individual liability; amendments in the society's rules and regulations require unanimous consent of the membership unless suitable provisions for modification have been provided for in the original scheme.

An incorporated body, on the other hand, becomes a corporation which, in many respects, is a 'person' in law with an independent continuous existence of its own, described as 'perpetual succession'.

The Companies Act of 1867, and a number of subsequent acts,

have provided for the formation of limited liability companies of the sort with which we are now all familiar. The liability of the members of a limited liability corporation is strictly limited and they are at risk only to the extent of their shareholding or guarantee. Corporations formed under the Companies Acts are required to put the word 'limited' after their name, but in the case of learned societies and qualifying associations there are arrangements whereby the Department of Trade and Industry allows this to be omitted.

Since 1867 many societies have been incorporated under the Companies Acts instead of by Royal Charter: for example, the Society for Analytical Chemistry, the Faraday Society, the Physical Society, and The Institute of Physics.

Incorporation as a limited liability company is simple, quick, and cheap, whereas to petition for, and obtain, a charter is rather more difficult and expensive. Even in 1847 it cost the Royal Society of Arts £400 in fees to obtain a charter and to-day the expense is of the order of several thousand pounds. Most of this is accounted for by the legal fees involved in preparing the petition and charter. The requirements for incorporation by Companies Act are simpler and involve a memorandum of association indicating the objects, and articles of association detailing the organization (Packman 1965).

Before the formation of the Institute of Chemistry, the R.S.A.'s activities embraced much applied science, including chemistry; for example, in the 1870s a special section was set up, dealing at first with applied chemistry and later with applied physics, and was extremely vigorous. In 1887 it was decided that the foundation of the Society of Chemical Industry and of the Institute of Chemistry made these special activities redundant, and they were given up.

Professionalization of the chemists can probably be regarded as being initiated by the Chemical Society in 1841 and being established by the formation of the Institute of Chemistry in 1877, when there was already a large and definable chemical industry.

Physics, or natural philosophy, on the other hand, did not pertain to a specific industry. Physicists were produced in universities, not in industry, as many chemists were, and the old universities had little enthusiasm for them.

It was not until 1874 that the Physical Society of London was formed with an initial membership of 80 (I.P.P.S. estimate, 1971, from the Minute Book), comprised almost entirely of university physicists. Physics had become a discipline, but not a profession.

By 1881 persons engaged on scientific pursuits had reached the total of 1200, and by 1891 some 2000. At the turn of the century the number was about 3200 and they comprised an important, though small, section of the community. Only 43 of this number were females.

The growing importance of physics was recognized by the formation of the National Physical Laboratory (N.P.L.) in 1902. It is of interest to note that the German counterpart, the Physikalische Technische Reichsanstalt, was established in 1887.

6

The early twentieth century and the First World War

THE later stages of the industrial revolution did, of course, necessitate the direct application of exact science, as in chemicals and electricity, but the real science-based industries emerged in the twentieth century, and the deliberate and continuous research for innovation after the first quarter of the present century. The industrial system slowly realized that it is dependent on science and technology and employed scientists and technologists in increasing numbers during the present century. The motivation was purely practical, to improve material and economic prosperity, and to increase profits.

The growth of the requirements for scientists is illustrated by the Census figures. By 1911 the number of persons engaged in scientific pursuits had risen to around 6400; only 145 of these were females.

By this time graduates from the science faculties of the new 'redbrick' universities were coming out in significant numbers and the old universities had adapted themselves to chemistry and physics. Nevertheless, apart from chemists few academically trained scientists were employed in industry.

What forces brought the university graduate into industry? What made the academic wet his feet?

The most disruptive influence on the academic life in the present century has been war. Modern war makes a large demand on the scientific and specialized talent of universities. Academic scientists were employed in great numbers in both world wars and made many notable contributions. Both wars forced university scientists into technology and industry, but the impact of academic chemistry occurred earlier than that of physics.

The First World War gave an enormous impetus to the chemical industry. It soon became evident that we had relied on Germany for many organic chemicals, particularly dyestuffs, and that our chemical industry was in no position to produce the numerous chemicals needed to wage war in the twentieth century. Vast quantities of various chemicals were required for manufacturing explosives and munitions of war, and in order to do this successfully many able men, including a large number of chemists, were brought into government service and obtained experience in large-scale chemical operations. The Government and nation began to recognize for the first time the importance of science and technology.

Before the end of the war a number of important steps had been taken towards creating the conditions for a British chemical industry. Between 1915 and 1919 the Government sponsored the formation of a modern dyestuffs industry in Britain with the heavy task of making up for all the years the locusts had eaten.

A further vital step in the development of the chemical industry, as we know it today, was taken by the British Government in 1917, when it decided to investigate the Haber–Bosch process. This initiated the modern high-pressure synthesis industry in Britain (Taylor 1961).

There was at this time a widespread determination that the country must continue its efforts to establish its own chemical industry independent of foreign countries. From this stems the immense progress and reorganization which occurred in the immediate postwar period. The prewar chemical firms, with a few exceptions, were small affairs, inadequately equipped technically and financially for the large-scale operations needed to establish vigorous and progressive units, but the war had thrown up many capable men and given them the knowledge and experience to innovate, consolidate, and advance, and the stage was set for a massive application of science, technology, and finance to the industry.

The First War period was characterized by an increasingly great participation by Government in scientific matters and the growth of the Scientific Civil Service. The Privy Council's Committee for Scientific and Industrial Research and an Advisory Council were formed in 1915, and the next year the Department of Scientific and Industrial Research was established with the task of encouraging research in universities and industry. It introduced a financial grant system to universities and students for research. This supported the new Ph.D. system which had been introduced as a reaction against the previously popular D.Phil. courses of the German universities, and thereby increased notably the supply of scientists. In 1916 the Advisory Council conceived the system of co-operative research in industry, which resulted in the formation of the current well-known research associations (Astbury 1968, Weck 1967). The D.S.I.R. organization lasted until 1964 when the Science and Technology Bill replaced it by the current organization.

By the end of the First World War the number of physicists had increased to the point when the urge for professionalization became strong. Discussions within the Physical Society led to an agreement in 1918 to form a new professional body, which was incorporated as the Institute of Physics (I.o.P.) in 1920 under the Companies Acts with licence to omit the 'Limited'. The new Institute became both a professional society and a qualifying association with its own examination system and grades of membership, designated by appropriate initials. The initial membership (1920) was 308. From its beginnings the I.o.P. contained a high proportion of physicists engaged in industry and in Government and research associations. In due course the ratio settled down to about 50 per cent academics and 50 per cent industrial members.

7 Between the wars and the Second World War

In the 1921 Census the number of people engaged in scientific pursuits was given as 12 400, of which over 700 were women. The numbers had doubled since 1911.

The growing importance of women in chemistry was recognized by the Chemical Society in its Humble Petition to George the Fifth in 1920 for a Supplemental Charter, as follows: 'The Membership of the Society has hitherto been restricted to males, but it is now desired, and is considered expedient for the welfare of the Society, that both males and females should be eligible for membership'.

In the postwar period scientists were recruited in increasing numbers by industry. There are two prerequisites for the effective use of the scientist and science in industry. First of all there must be a climate of opinion within industry which accepts the scientist on equal terms with others, and secondly there must be the acceptance by the scientist of the industry in which he works and an understanding of, and identification with, its objects.

Traditionally, the industries of Britain had been headed by men of affairs and social position, rather than by specialists or scientists. There were exceptions to this, notably in engineering, where the prestige of the professional had been established and the engineer himself had been long enough in industry to come to terms with his environment and apply his disciplines to the whole range of activities, in a rounded manner. His was no ivory tower attitude, and he had no prejudice or suspicion that trade and commerce were vulgar and unattractive.

Identification with an industrial environment is not natural to the scientist. The marshalling of scientific principles for utilitarian ends is not an objective normally cultivated in university pure science courses, and it may take a long time to acquire the faculty by personal experience. The scientist has been trained in an intellectual atmosphere, which even now to some degree considers application of research as a prostitution. I can well remember when I was at university that there was among many of the academic scientists, including, of course, the 'angry young men', a contempt for the utilitarian application of science and a reluctance to participate in industry. This arose, not only from youth and inexperience, but from the profound conviction that the object of science was the exploration of knowledge for its own sake, and nothing else. In retrospect, it is difficult to understand such an arrogant attitude, but I appreciate it, since I shared it. It was broken down by the economic conditions in the late twenties and early thirties which forced young scientists, who like other people must eat, to earn their livings where they could. More important still, the attitude was changing because of the transformation which was taking place in industry itself.

Laboratories, facilities, and conditions in industry, which had compared very unfavourably with those in universities, were improving rapidly owing to the influence of scientists and were beginning to attract recruits.

Physics was in a transitional period, in process of extending seriously and deliberately into technology, and this was well seen in the Cavendish Laboratory where the 'string and sealing-wax' methods, which had been so adequate and rewarding during the previous two decades, were giving place to techniques derived from the engineering and electrical industries. These were particularly evident in the work of Kapitza and of Cockcroft and Walton, to quote but two examples in the Cavendish Laboratory. The technological impetus of the First World War had established the position of physics in the electrical industry, and the great lamp works, such as G.E.C. and Phillips, were employing physicists in considerable numbers and making great strides in vacuum and other techniques. Products such as thermionic valves and X-ray tubes were being 'fed back' to academic physics. In the same industries physical chemistry was coming into its own, having established its worth in the work of people like Langmuir. A scientific instrument industry was emerging and the oil industry was employing more and more physicists for seismographic prospecting.

In the chemical industry economic and technical forces had resulted in the formation of a number of large, powerful, and scientifically competent companies by integration, particularly in Germany and the United States. The synthesis of ammonia, pursued so efficiently by Badische at Oppau under the impetus of war, had realized man's dream of abundant fertilizers from the air, and the modern synthesis of chemicals had really started with all its complicated techniques.

By 1931 the personnel engaged on scientific work had increased to about 19 000, of which over 1000 were females. The population explosion of the scientific community was beginning.

The inter-war years saw the consolidation of the science-based chemical and electrical industries and the period of deliberate and continuous research for innovation. It saw the beginnings of vast developments in the plastics and man-made fibre industries with such notable products as polyethylene and nylon. It also saw the emergence once more of the chemist-businessman and the scientist-businessman as important components in modern industry. In some companies, for example, men of scientific training were widely distributed in the financial, commercial, sales, and personnel departments. This has had the inevitable result that the scientific approach and methods have become increasingly applied to all the activities of the industries.

In the Second World War the whole scientific and technological resources and personnel of the nation were mobilized. Notable advances in electronics, aerospace activities, and medicinals stemmed from this effort and once more academics were involved in large numbers. Moreover, many techniques of decision-making, such as operational research and analysis of requirements, were evolved by scientists during this period.

The Kaiser's war has often been described as the chemists' war and Hitler's war as the physicists' war. It is probably more accurate to describe the First World War as dominated by chemistry and engineering, whereas the Second World War was fundamentally one of science (particularly physics) and technology. It was the Second World War which made the marriage of science and technology inevitable.

The 'scientific population', which was about 19 000 in 1931, had risen to over 44 000 in 1951, and of these about 3500 were females. Science had become institutionalized.

8

The institutionalization of science

WE saw in Chapter 4 how the professional aspirations, which were characteristic of the Victorian middle classes, resulted in the formation of engineering and scientific professional and qualifying associations. This was but one manifestation of a world trend to institutionalization of science in its role in society. Shils (1968) has described the institutionalization of the scientific role and the establishment of science as a profession in a wider context than that dealt with in Chapter 4. 'The major activities of science — the teaching of science and scientific research — are usually nowadays services which the scientist offers in exchange for his livelihood.' The great increase in these activities resulted in the gradual institutionalization of science.

Scientific identity developed early in England in the late seventeenth and early eighteenth centuries, but the practice of science as a recognized profession lagged behind. According to Shils, Germany was probably the first country to recognize science as a profession, whereas Great Britain was one of the last to achieve this state in the late nineteenth century.

The process of institutionalization has accelerated during the present century because of the importance of science in national defence and economics, and has been brought about to a large extent by Government assistance, intervention, and action in both education and research. During the present century many institutions have been established and modified to form the framework of this modern role of science. The main functions of these establishments are to educate and train personnel for the scientific profession and to provide continuity for its activities. There is also a growing requirement for supervising, controlling, and advising on scientific policies and expenditure on a national scale; this role can be defined as 'the government of science'.

The universities and educational establishments have played, and still are playing, a large part in the institutionalization process.

When the physical sciences developed as separate disciplines in Britain during the nineteenth century, they were regarded as alien by the old universities, which were still dominated by clerics in Holy Orders, as they were in the Middle Ages. Higher learning in the universities was supposed to be the pursuit of truth. This, as we saw earlier, was the medieval scholastic innovation adopted during the Renaissance and in essence a hybrid of Greek classical philosophy and the Christian pursuit of the Will of God.

The sciences could be made respectable only by overlaying them with this traditional motivation, and so scientific studies also became part of the pursuit of truth. We have paid a high price for this aberration which has been responsible for the present dichotomy and fragmentation of science and technology.

After the founding of colleges of science and technology in the middle of the nineteenth century, and later, and the growth in size and numbers of the redbrick universities in the present century, the institutionalization of science accelerated notably.

After the Second World War the universities, only recently attuned to science, were confronted with the task of adapting themselves to technology, and this painful exercise has continued from then onwards. Since the Second World War the exercise has assumed massive proportions and is still under earnest consideration.

In 1964 the Council for National Academic Awards (C.N.A.A.) was set up to provide degrees outside the normal university system, and successively technical colleges have been built up for this new role of science and technology and many transformed later into colleges of advanced technology (C.A.T.s.) and then into universities. The functions of polytechnics and colleges of education have also been expanded and modified (Wood 1968, p. 65).

Government expenditures in connection with universities are currently dispensed and monitored by the University Grants Committee (U.G.C.) (Wood 1968, p. 67). The Secretary of State for Education and Science is responsible for disbursements for education and that part of science which may roughly be termed non-applied. The Secretary of State for Trade and Industry is responsible for applied science and technology and industry. The Secretary of State for Defence is responsible for R and D relating to the armed forces. In addition, the Prime Minister has a chief scientific adviser and supporting scientific staff in the Cabinet Office and a Central Policy Review Staff under a Director-General (currently Lord Rothschild). The Government in respect of scientific matters normally looks for advice to the Royal Society, which is the body of paramount status in the scientific world. It receives an annual grant from the D.E.S. (currently about £0.9 million). With Royal Society assistance the Government has set up various bodies to advise on its expenditure on science and to monitor results. Collectively these bodies constitute an important factor in the institutionalization process. A Council for Scientific Policy (C.S.P.) was set up in January 1965 'to advise the Secretary of State for Education and Science in the exercise of his responsibility for the formation and execution of Government science policy'. The Council has 15 members drawn from universities, industry, and research establishments. The C.S.P. organization has five Research Councils which are, themselves, autonomous, incorporated by Royal Charter and receiving grants in aid: these are the Agricultural Research Council (A.R.C.) founded in 1931; the Medical Research Council (M.R.C.) originally set up in 1913; the National Environment Research Council (N.E.R.C.); the Science Research Council (S.R.C.), and the Social Science Research Council (S.S.R.C.), the last three set up in 1965.

The central Government is at present reviewing the functions of its various departments and endeavouring to 'provide a framework within which the Government's policies, as a whole, could be more effectively formulated'. To this end the Central Policy Review Staff, frequently referred to as the Government 'think-tank', was set up.

The Government commissioned a Report from the Director-General, Lord Rothschild, on R and D activities. As a result the Green Paper *A Framework for Government Research and Development* (HMSO 1971, Cmnd. 4814) was produced. This comprises two main portions, 'The organisation and management of Government R and D' by Lord Rothschild, and 'The future of the Research Council System', which is a report of a C.S.P. Working Group under the chairmanship of Sir Frederick Dainton. This gives an up-to-date description of the Council for Scientific Policy (C.S.P.) and the present Research Council system and its finances. Both sections of Cmnd. 4814 make recommendations for organizational changes, which are the centre of intense discussion and disagreement, particularly the Rothschild customer–contractor principle.

An important feature in the institutionalization of science in Britain has been the research establishments sponsored by the D.S.I.R. which was set up in 1918 (Weck 1967, Astbury 1968). These comprised Government research stations carrying out work fully financed by Government funds, and research associations partially financed by participating firms, in particular firms in competitive industry. The latter are a very particular feature of British co-operative research.

Since the termination of the D.S.I.R. in 1964, responsibility for the old D.S.I.R. research stations has been transferred to various Government departments. The number of these research stations carrying out work fully financed by Government funds is currently (1972) 15, but there are, in addition, scores of other Government research establishments, for example in the Ministry of Defence.

The number of research associations has been slowly declining and is now 40.

9

Scientists and technologists

The institutionalization of science has had far-reaching effects. According to Shils (1968), 'The institutionalization of science meant that science came to be carried on as a discipline, or, rather, as a set of numerous interconnected subdisciplines, each one of which appeared to its practitioners as more or less self-sufficient'. It resulted in intense specialization; in the artificial separation into pure and applied science and technology and a further separation from engineering. This was accompanied by a growing demand for, and a concentration on, academic or equivalent qualifications which displaced the old system of apprenticeship and tutelage. Statistics now concentrate on 'persons holding degree or equivalent qualifications in engineering, technology and science'.

The fact that the new and, to a large extent, the present generation of technologists are mainly university-trained has induced profound changes. We are apt to forget that the majority of the inventions and discoveries which were the basis of British prosperity for a period of about 75 years were not produced in universities, nor were the personnel who ran our successful industries. The old generation came from inside industry and understood it, whereas the new generation comes from outside industry, and there is still a great deal of mutual distrust and misunderstanding. So far, too, the pursuit of science for utilitarian ends via industry has not been accorded the status and importance it has in other countries.

The position is now changing, but until recently it could be said that in Britain the scientist had status, but little power, whereas the businessman had power, but little status; the technologist had neither and was regarded too often as a second-class citizen in the world of science. This has not been the case in other great industrial societies. The scientist–businessman played a great part in Germany in building up industries, such as the chemical industry, and its scientists, technologists, and businessmen have always had status and position. In the USSR also they have acquired high position and, above all, in the USA there is prestige and influence in all these occupations.

It is only recently that a general awareness is arising that as a nation Britain is entirely dependent on industry to survive in international competition and that British society is based on technology. Hitherto, industry, trade and technology have, to no small extent, been regarded as second-class employments, but recent Governments have also become aware of the facts of life, recognizing that industry and technology must be encouraged and given status and prestige.

It is one thing to recognize a situation and quite another to rectify it. It is important, however, to examine the position of scientists and technologists in Britain in relation to their status in the community.

The scientist has achieved status through the Royal Society. He has also, after a considerable struggle, become established in the universities and shares in their status and prestige. Technologists have not been so fortunate.

The dichotomy of science into pure and applied, of comparatively recent origin, has resulted in a preponderance of academics in the Royal Society's fellowship which includes few technologists. As a result, technology has not enjoyed the status that pure science has acquired from its identification with the Royal Society.

The position was stated clearly in a letter from Sir Gordon Sutherland published in the *Guardian* (30 November 1963), and I quote from this:

> The importance of having a large increase in the number of young people with first-class intellects taking up technology as a career has been stressed by innumerable people in the last few years . . . A recent survey showed that the proportion of able youngsters in the sixth forms opting for a training in engineering, or technology, is very low indeed. What can be done to change this situation? The basic trouble is that in the schools and universities engineering and technology are accorded a second-class rating. The result is that young people become imbued with the idea that 'pure' science and 'pure' research leading to Fellowships in the Royal Society are the ideals

towards which they should strive. There is no doubt that the Royal Society has had a profound influence in placing and keeping Britain in the forefront of pure science.

Sutherland went on to propose that a new society, a sort of Royal Society for Technology, should be formed which would in due course confer status and standing on the élite of the technological world. Such a society has not yet emerged, but there appears to be a wind of change blowing through the Royal Society. Sir John Wolfenden, in his speech at its Anniversary Dinner in 1964, remarked: 'I venture to suggest to you that those words "useful arts" might well, in the year 1964, be interpreted as being something to do with activities of a scientific kind which are not necessarily what are, or have been, dignified in the past with the adjective "pure"'. Sir Howard Florey, the President, in reply, said: 'Some of our friends might like to know that we are hoping to augment our strength and our standing in the community by increasing the number of applied scientists among our Fellowship and, perhaps, eventually, to widen our intake of Fellows by including some whose principal interests are the study of man himself'. Indeed, recent elections confirm the intention of the Royal Society to increase the number of applied scientists among its Fellowship.

There is, however, a view, which has much merit, that there is a paramount need for an independent body of high status and integrity to maintain and further the pursuit of pure scientific knowledge, that the Royal Society during the last century has achieved this position, and that it should be maintained intact as such. The proponents of this view would argue that the Society should not be 'diluted' by technologists and engineers, for this would inevitably lead to a concentration on the useful rather than the fundamental.

It would be possible to form a technology body on the lines of the Swedish Royal Academy of Engineering Sciences, but it could hardly have the prestige of the Royal Society. The American parallel is worth mentioning.

In early 1960, American engineers were complaining about their professional status and their public image, and the idea was conceived of forming a National Academy of Engineering. The President of the National Academy of Sciences, which is the counterpart of the R.S., had the bright idea that as the N.A.S. charter contained provisions for forming an academy within an academy, the Engineering Academy could operate under its charter (Walker 1969). Recently the Academy of Engineering appears to have run into some difficulties.

It is unusual, if not unique, for charters to allow for academies within academies. It is considered by legal opinion not to be feasible in the case of royal charters. In this respect it may be observed that the R.S.A. in recent years has considered this matter of creating a body for special recognition of outstanding technologists and industrialists. It concluded that it was not appropriate for it to initiate such a project, although it was fully aware of its feasibility from a previous exercise when it wished to recognize outstanding designers.

To do this the R.S.A. Council had made an Ordinance in 1937 establishing (after approval by the Crown) a Faculty of Designers for Industry of which each constituent member would have the title 'Royal Designer for Industry', and use the letters R.D.I. after his name. The membership was limited originally to forty, and the persons chosen had attained eminence and efficiency in creative design of a high aesthetic standard for industry. It is to be presumed that other chartered bodies could create similar faculties if they so wished.

It will be realized that the Royal Society and the academics can play a great part, if they so desire, in establishing the status of technology in this country. While the Royal Society can do much to change the climate of opinion, real progress can come only from the grass roots, the schools and universities.

Thring (1965), in an article on 'The nature of pure and applied physics', wrote:

> Britain can no longer afford to teach many of the best physics students the false idea that pure physics offers a superior intellectual satisfaction or a more honourable career than applied physics. The truth is that they are indistinguishable as regards quality of thought, and depth of scientific penetrations: they only differ in aim and in choice of field of interest to that aim.

Everyone can appreciate the fascination of carrying out pure research and the absolute necessity of promoting it, and it is only the cult of exclusivity to which we can rightly object.

Thomson, in his Presidential Address to the British Association in 1960, summed up the situation as follows: 'Science is not merely the control, but also the understanding of nature. Its two aspects must be held in equal honour'. Unfortunately, it takes a long time for top opinion and views to seep down to the schoolmasters and university teachers, and the educated public, who were raised in the old views and with the old prejudices. This is our problem, and we shall have to find accelerated means of re-educating those who are responsible for forming the views and climate of opinion of the young people with whom the future rests. If we do not, there will be a continuing belief in the young that to become technologists is, at best, a second choice, rather than a positive preference.

Since the war vast science-based industries, such as pharmaceuticals, electronics, computers, and the aerospace industry, have burgeoned. These require large numbers of scientists, both for manufacture and research and development. Concurrently with these developments, research has developed in such a way that it is more and more dependent on highly sophisticated technology, for example space research, using the Jodrell Bank telescope; physical research employing million-volt electron microscopes, or the linear accelerators. Research scientists, both in universities and industry, are thus involved more and more in technology which often is the predominating element in any piece of research. Moreover, since advanced technology is extremely expensive, scientists are becoming increasingly concerned with administration, finance, value analysis, and the like.

10

The scientific community today

THE scientific community has increased exponentially from less than 500 in 1851 to around 175 000 in 1966. During this time the population has grown from around 20 million to about 50 million.

Information on the current scientific community is included in the 1966 Sample Census of Great Britain, particulars of persons with scientific and technological qualifications being given in a Summary Document (H.M.S.O. 1971). The total number of people qualified in science or engineering and technology was 356 760 (318 640 males and 38 120 females). Of these, 174 990 were qualified in science (137 740 males, 37 250 females). On this basis of definition the scientific community of the country was around 175 000 in 1966.

Roughly 75 000 of the scientific community are engaged in industry, trade, and commerce; 54 000 in education; and about 10 000 in scientific and technical services, research and development, and scientific bodies. Of these 10 000 it is probable that about half are employed on behalf of industry; of those in industry, 4510 are described as managers (including a fair sprinkling of sales managers). Of the 54 000 engaged in educational services, about 30 000 are in schools and 23 000 in establishments for further education.

Further, it is to be noted that there are roughly 134 000 males and 37 000 females in the scientific community, a ratio of 4 to 1. Of the females, about 14 000 are in the educational services, which is more than half of the females economically active (about 22 000). Of the total scientifically qualified females (37 000), 15 600 are economically inactive, presumably due to their marital state.

Over 50 per cent of the scientific community are under 39, and over 30 per cent are under 29; 12 per cent are over 60 years of age.

There are 69 280 persons with post-graduate qualifications in science, and of these 11 050 are classed as scientists and 32 000 as teachers. There are 18 510 persons classified as having Ph.D., D.Phil., D.Sc., or D.V.Sc. degrees.

Statistics of Science and Technology (H.M.S.O., 1970), issued by the Department of Education and Science and the Ministry of Technology, contains interesting information on expenditure on R and D in the United Kingdom, on scientific, engineering, and technological manpower, and on students in educational establishments. The manpower surveys differ from the census of population in that they omit certain categories (see p. 109 of *Statistics of Science and Technology*, 1970).

In 1968 the publication listed about a quarter of a million persons holding degree, or equivalent, qualifications in engineering, science, and technology. Of these, 102 000 were qualified in science. Of the 102 000 qualified scientists listed for 1968, about half were employed in education, some 15 per cent being in universities, half of these in pure science. Some 40 per cent were engaged in industry and about 12 per cent in government service. The percentage engaged in R and D (excluding education) is given as somewhat more than 50 per cent, but the bulk of this is applied research and development.

Science is now professionalized: the day of the gifted amateur is over. Most scientists are employees, and currently only about 1 per cent are self-employed.

It seems probable from the information available that about 10 per cent of Britain's total R and D effort is on basic research defined as 'work undertaken for the advancement of knowledge without specific commercial objectives', and that about 2½ per cent of British scientists are engaged on such work.

The scientific community as it is organized today is engaged on research, instruction, and application of its scientific knowledge. It provides a service for our science-based industrial society for which its scientific education and methods peculiarly fit it. To this end it is involved in technology and engineering and it operates in universities, educational establishments, Government (both national and local) and nationalized bodies and in industry, commerce, and trade.

It is clear from this analysis that the scientific community is spread throughout all the nation's activities and that the educational bodies and the scientific societies and institutions which serve this community must cater, not only for the academics, but also for the very much larger numbers in industry, trade, and commerce.

Cotgrove and Box (1970) have carried out a survey of scientists in the universities and in the chemical industry, their stresses and motivation. They identify a typology of scientists corresponding to the academic, application, and organizational roles in science as follows:

Public scientists, who maintain the academic tradition and see their rewards by personal reputation via publication of original work. *Private* scientists, who may be regarded as non-academic professionals quite willing and competent 'in exploring the problems of relating science to industry' without publication. Such men achieve rewards by virtue of their status and position in their own private industrial organizations. A third category is the *organizational* scientist, or administrator. Such exist in both the academic and industrial worlds.

If we are concerned with the whole of the scientific community, as we are, a fourth category is identifiable, namely, that important section engaged in educational work who obtain their reward by undertaking the task of training the next generation of scientists.

The conclusion is inescapable that the scientific community is using 'science' primarily as a means of gaining a livelihood, and the sooner we recognize fact and fiction and abandon the academic myth the better. It is also becoming increasingly clear that it will not be possible to continue an exponential rise in the scientific community and provide them all with 'scientific' occupations. Scientific education in the future will have to be planned as an alternative to the classical one as an entry to a wide spectrum of jobs. As in classics, there will be scholars and philosophers, both experimental and theoretical, to advance knowledge and maintain standards of excellence, but that will not be the sole, or even the main, purpose of scientific education.

11

Scientific societies in a changing world

THE great increase in the scientific population brought new problems, more particularly those of organization of their concerted activities in societies and institutes.

Small societies, comprising not more than a hundred or so members, can be, and were, run by honorary officers who give their services free, supported by perhaps one full-time, or part-time, paid secretary. They were frequently related to a university department, or to some organization which supplied some services and carried the overheads. As societies become larger, the range of activities increases, in meetings, in publications, and so on, and it is no longer feasible to run them on voluntary effort alone. Full-time staff and more complex organization are needed, and the bodies end by having to carry the full commercial costs of their activities on their own finances, or else by Government or industrial-company subsidies.

Since the war, continuous, and in recent periods galloping, inflation has greatly affected learned societies, whose activities are labour-intensive, providing personal services. They have for many years faced rising costs, which are not matched by proportional rises in receipts and which cannot, as a rule, be recouped by increasing members' dues. Many bodies are in financial difficulties, or precariously balanced on a knife-edge, and depend on surpluses from their publishing or examination activities, which themselves are subject to strain and erosion.

A number of societies have mounted appeals to industry to strengthen and fortify their finances. There was a period after the war when industry was expanding and profitable, when it could afford to be, and was generous, as a patron of science. Appeals by the Royal Society, the Society of the Chemical Industry, the Institute of Physics and the Physical Society, the Chemical Society, the British Association and the Royal Institution brought in considerable funds. Since the early sixties the climate has changed and it is now difficult to obtain funds from private firms for the general objectives of learned societies. Funds from Government are equally hard to come by. Most societies consequently have to rely on their own efforts and most have had to advance membership fees. There is, however, a limit to what can be raised in this manner, because the scientific community, in common with others, is feeling the pinch. Moreover, members of scientific bodies are increasingly asking what they are getting out of their membership fees.

The growing recognition of the unity of science and technology and the great increase in university graduates in the scientific community in the fifties turned scientists' thoughts to possibilities of co-operation and merging of interests in the learned societies and professional bodies. The basic reason was probably the wish to avoid duplication of effort and to provide better facilities and service, but there was also the thought of bringing the academics and industrialists together and helping to reduce the dichotomy of science and technology. This initiated a process which began to reverse the division between learned societies and professional and qualifying associations which had taken place at the end of the nineteenth century and the beginning of the twentieth century.

Earlier mergers were not unknown, but were unusual. For example, in 1927 the Royal Aeronautical Society incorporated the Institution of Aeronautical Engineers, and in 1945 the Institution of Mechanical Engineers was joined by the Institution of Automobile Engineers.

There are a number of societies which provide for kindred societies or industrial concerns to participate in some formal fashion in their activities. Thus, the R.S.A. has 'Institutions in Union', which include kindred societies, libraries, colleges, schools and research organizations. The R.S.A. has, additionally, 'Companies in Association', comprising limited liability companies or other trading bodies. The B.A. has a category designated 'Corresponding Societies', comprising societies with similar or complementary aims. The R.I. has a category of 'Corporate Subscribers', which includes any corporate or incorporated body accepted by the Committee of Managers.

Such schemes are most useful in promoting communication and collaboration between organizations and could be more widely adopted. They do not, however, provide for the closer links which many believe to be necessary in the world of today.

There had been efforts to integrate the activities of the chemical bodies, for example the Chemical Council (1935), and its expansion into the Conjoint Office (1941), with a joint subscription system, but they had no permanent success. Probably the climate of opinion at the time was not favourable to such mergers and no compelling factors had arisen to make them inevitable.

Opinion in favour of joint action was steadily mounting in the 1950s, and joint meetings and common activities were arranged by many scientific and technological bodies. For example, the Chemical Society and the R.I.C. had adopted a common house journal and a common A.G.M. and annual meeting time, but this was only a beginning.

12
The physicists unite and the professional institutes get together

AMONG the scientists, the physicists took the initiative in merging, and the Physical Society and the Institute of Physics commenced discussions for closer collaboration in 1957. Both bodies were limited liability companies and both were put into liquidation after agreement had been reached to merge the two bodies in 1960. The new body was called the 'The Institute of Physics and the Physical Society' (I.P.P.S.) and was incorporated as a company limited by guarantee and without share capital under the Companies Act 1948. The objects were 'to promote the advancement and dissemination of a knowledge of physics, pure and applied, and the elevation of the profession of physics'. In a manner of speaking, this was a pioneering effort in that for the first time a learned society, which was a charity in law, was joined to a qualifying association which did not rank as a charity. The charitable status of the learned society and the educational activities of both bodies were maintained via the Physics Trust Fund, the sole objects of which were to promote the advancement and dissemination of a knowledge of physics, pure and applied, but not the elevation of the profession.

The membership of the Institute in June 1960, before amalgamation, was 6958 and of the I.P.P.S. in December 1960, after the amalgamation, 8720. By 1965 the numbers were 11 273 and in 1970, 15 926. It is believed that about 50 per cent of the physicists in the U.K. were members of the I.P.P.S. and, of the total membership, about half were engaged in industry. Membership had grown from some 300 in 1920 to 16 000 in 1970, which gives some indication of the very rapid growth of the physicist population since the First World War.

The next important move towards collaboration was not in the scientific bodies but in engineering. This move was of importance to the scientific community, not only because of its close connections with engineering, but because many scientists are members of engineering institutes.

Engineers, like other technologists, had long considered that they lacked status and representation in the affairs of the nation in which they played such an important part. It was therefore agreed to establish a National Association of Engineering Institutions so that they could act in concert in matters of mutual interest.

A draft constitution for the new British engineering body was adopted at the inaugural meeting in October 1962, and it was named 'The Engineering Institution Joint Council'. The initial Constituent Members of the Council were: the Royal Aeronautical Society, The Institution of Chemical Engineers, The Institution of Civil Engineers, The Institution of Electrical Engineers, The Institution of Electronic and Radio Engineers, The Institution of Gas Engineers, The Institute of Marine Engineers, The Institution of Mechanical Engineers, The Institution of Mining Engineers, The Institution of Mining and Metallurgy, The Institution of Municipal Engineers, The Institution of Production Engineers, and The Institution of Structural Engineers.

The original body was informal and unincorporated, but it proceeded with considerable speed to petition for a royal charter and was incorporated in 1967, with the title 'The Council of Engineering Institutions' (C.E.I.). The constituent members all retain their independence and sovereignty, but act in concert in relation to matters of common interest, such as standards of examination, general policy and representation of views. The combined memberships carry the title of 'Chartered Engineers' (C. Eng.). The C.E.I. has become increasingly active in propagating the views of the engineering community and serving as a mouthpiece and pressure group on matters of general interest and importance to its constituents.

A further step in collaboration between technological and scientific bodies was initiated in early 1967 by Mr. J. A. Ratcliffe, then President of the Institution of Electrical Engineers, and a past president of the Physical Society. Meetings were held between the Institution of Electrical Engineers, the Institution of Electronic and Radio Engineers, the Institute of Mathematics and its Applications, and the Institute of Physics and the Physical Society. This led to the formation (in June 1968) of an informal body called the 'Standing Committee of Kindred Societies to co-ordinate Activities of Mutual Interest'.

In 1967, shortly before the C.E.I. received its royal charter, the Chairman of the Council of Engineering Institutions, together with some presidents of member institutions, met with some presidents of science institutions to discuss areas of common interest. This resulted early in 1968 in the formation of a 'Science/Engineering Institutions Joint Affairs Committee', comprising C.E.I. and science institutes representatives (R.I.C., Institute of Metallurgists, I.P.P.S., Institutes of Biology, of Fuel, and of Mathematics and its Applications).

The possibility of forming a single body embracing both C.E.I. and the science institutes was discussed, but the general view was that this was not practicable unless the science institutes first formed a federation of their own so that they could speak with one voice.

Discussions were undertaken in July 1968 on proposals for a Federation of Science Institutes to promote common objectives and it was decided to proceed with the project, the name finally chosen being 'The Council of Science and Technology Institutes' (C.S.T.I.). C.S.T.I. was established as an unofficial body as from 1 January 1969.

The total membership of the C.S.T.I. at its formation amounted to approximately 58 000, made up as follows: Royal Institute of Chemistry, 25 000; the Institute of Physics and the Physical Society, 14 000; Institution of Metallurgists, 10 000; Institute of Biology, 6000; Institute of Mathematics and its Application, 3000. Subsequently (1971), the C.S.T.I. applied for incorporation as a company under the Companies Act 1948.

Meanwhile, the physicists were also busy in other directions. From 1967 they took an impressive lead in bringing into being the European Physical Society (E.P.S.), which was incorporated in September 1968 in Geneva. The I.P.P.S. membership was brought into the E.P.S. as a body, but there was provision for individual membership as well. Thus a European Community of Physicists was initiated, of which the I.P.P.S. comprised the greatest number. This may well be of increasing importance when Britain joins the E.E.C.

In 1967 the I.P.P.S. displayed a further imaginative initiative in purchasing the Fulmer Research Institute. The Institute is a company which carries out sponsored research work on a commercial basis. This is the first time that a learned society has acquired research laboratories for doing sponsored work. So far the experiment has been successful and profitable.

In 1966–7, the I.P.P.S. began to consider the possibility of petitioning for a royal charter. The merger had settled down by this time and it was considered advisable to establish the I.P.P.S. as the national body primarily concerned with physics with the influence and status accorded by a charter. This was a formidable exercise which owed much to the persistence, hard work, and competence of the Honorary Secretary, Dr. Robert Press.

It was inevitable that some compromises had to be accepted and some sacrifices made to achieve a charter which could encompass in a single body both the learned society and the professional and qualifying activities of the I.P.P.S. The Privy Council, which examines petitions for charters with great care, has certain standards which it has adopted after centuries of experience. Since a charter grants perpetual succession for the body to carry out the objects defined in its charter, the Privy Council does not look favourably at composite titles which might suggest a subsequent schism or separation. For this reason it was necessary to drop the dual title and adopt a single title which was finally chosen, because of the qualifying functions, as 'The Institute of Physics'. Furthermore, it was unfortunately necessary to drop the designation 'Fellow of the Physical Society' as inappropriate in a qualifying association since it does not represent a 'qualified' grade.

There were, not unnaturally, some people who were opposed to these changes, but the bulk of the membership was in favour of the proposals for a charter which were put to them at a number of meetings. The proposals were approved and a petition for a charter was made to the Sovereign and granted on 6 November 1971.

13

The chemists unite

THE views in favour of merging of the chemical bodies were given in a letter which I sent in 1967 to *Chemistry and Industry* (Taylor 1967):

As one who has a foot in many camps I am writing this letter on a subject which is of current interest to many chemists.

Having spent my working life in the chemical industry I am a Member of your Society, a Member of the Chemical Society and a Fellow of The Royal Institute of Chemistry, but, having been trained initially as a physicist, I am also a Fellow of The Institute of Physics. As an Honorary Officer of The Institute of Physics, I took part in the negotiations between the Institute of Physics and The Physical Society, which resulted in the amalgamation of the two Bodies in 1960. The present 'Institute of Physics and The Physical Society' has thus operated for six years and, during this period, we have seen the emergence of a unified body with a singleness of purpose and an increased capacity to cater for the requirements of Physics and Physicists in this modern technological age.

My own personal experience of this amalgamation prompts me to suggest that it is high time that chemists undertake a similar course of action. I am well aware of the difficulties involved, but I do not believe any of them are insuperable, provided the negotiators are clear as to their objectives; understand that they are arranging 'mergers' (sometimes referred to as 'marriages of equals'), not 'take-overs', and are prepared to make generous compromises to achieve their ends.

The Membership, as well as the Councils, of the negotiating organizations must, of course, be convinced that amalgamation is in their best interests.

It may be useful, therefore, to recapitulate the principal reasons for amalgamation which are, I believe:

(I) If scientists, technologists, and others employed in the science-based industries, are to make their presence felt in our technologically-based society and have the influence to which they are entitled, they will have to present a united front. Fragmentation of the chemical population merely results in the weakening of its influence and status. Unity is strength.

(2) If Britain enters the Common Market, as she must, sooner or later, it will be an enormous advantage if a single powerful body can speak for chemistry; to seize the great opportunities which will be presented, and to deal with the many delicate situations which will arise.

(3) In the present state of the economy it is becoming increasingly difficult to finance the activities of Learned Societies. Responses to Appeals for Funds are meagre in the present economic climate and it would not be easy for Members to find the money to pay the greatly increased Annual Subscriptions which would be justified in view of the increases in costs which have taken place in recent years. It is essential to try to limit the costs of the permanent staff and of accommodation and one way of doing this, whilst maintaining, or even improving, the standard of service to Members, is by amalgamation with similar bodies so that overhead expenses can be shared. This also has the advantage that the financial strain on individuals in having to belong to a number of bodies is reduced, even if the Subscription for the joint body is increased.

(4) There is currently much overlapping in the Meetings, Symposia, Conferences and Publications of bodies dealing with similar scientific and technical subject matter. It is appreciated that much has been done in the form of joint meetings, etc., to ameliorate this position, but the simplest method is to have one organization. I believe this is particularly important for the Branch, Section and Group activities, especially those of the Provinces.

(5) Publications are an important activity of most organizations and they are so costly and vital financially that they ought to be run as businesses. Learned Publications, etc., are subject to increasing competition from Commercial Publications. Most scientists believe, nevertheless, that it would be a great loss if the Scientific Societies were not able to continue as independent Publishers with their own policies, standards and independence. It is a considerable advantage economically to have larger publishing activities and this can be achieved by amalgamation.

(6) There is a further reason in favour of mergers, which is usually overlooked, but which I believe is important. Whilst the permanent staff are competent and devoted, they are very thin on the ground and Scientific Bodies could not function at all but

for the enthusiastic and unpaid work of the Honorary Officers. This is true not only of the 'Headquarters' Honorary Officers, but also of those in the Branches and Groups. The supply of men, capable and willing to take on such onerous unpaid duties, is limited and, under modern conditions they are very extended: indeed, it is often the same few people who carry the burdens of the various Institutions and Societies. Mergers would materially conserve this great source of initiative and energy.

The SCI membership embraces people from many disciplines and professions, ranging from the Law to Publicity, all having a common interest in the chemical industry and, in my view, it is most important that they should be in a position collectively to make a full contribution in the matters which I have enumerated above.

Many further reasons could be adduced in favour of merging, but I think that those I have given provide a strong prima facie case for the various Chemical Bodies to make a serious and open-minded study of the possibilities.

The aspirations of chemists for amalgamation were brought into the open by Sir Harry Melville in his Presidential Address to the Chemical Society in April 1967, when he called for a closer association between the various societies covering the different aspects of chemistry. After this, informal discussion took place between the Presidents of the Chemical Society, the R.I.C., and the S.C.I., with the result that a joint Official Notice from the Councils of these three chartered chemical bodies appeared in the house journals in late 1967, indicating that a study was to be made under the guidance of Sir Eric Bingen on the possibilities of amalgamation.

In the event, Sir Eric was unable to carry out the task because of ill health. In consequence, the writer was asked to undertake the investigation.

In May 1968 the writer submitted a report to the Presidents of the three chemical bodies under the title 'Examination of Possible Methods of Merging the three Chemical Bodies'. This was a critical study of the various possibilities, the advantages and disadvantages, and so on, not a recommendation for a final plan. A negotiating committee was then set up by the three bodies, of which the writer was invited to be chairman. The report formed the basis for subsequent discussions. These were difficult and protracted, but they were characterized by great patience and persistence on the part of the negotiators, who were determined to succeed in their object.

Initial discussions were between the C.S., the R.I.C., and the S.C.I., but in late 1969 the Faraday Society (F.S.) and the Society for Analytical Chemistry (S.A.C.) joined in. By early 1969 the C.S. and R.I.C. representatives were satisfied that a scheme for merging their common interests could be agreed, but the S.C.I. negotiators faced considerable difficulties. A three-body scheme not being practicable, the C.S. and R.I.C. proceeded to negotiate one for their two bodies, with provision for participation of others, should they so desire. This scheme was developed and published in October 1969 (Williams, Jones and Taylor 1969). It was modified somewhat after later discussions with the F.S. and S.A.C.

The memberships of the various bodies were kept informed of the progress of the negotiations and the final scheme was made the subject of referenda by the individual memberships, who expressed their approval. The four bodies (the C.S., R.I.C., F.S., and S.A.C.), then produced a document, 'The Amalgamation Proposals', which was approved by appropriate resolutions at general meetings of the respective bodies. The scheme came into operation on the appointed date of 1 January 1972 (Porter 1971).

The estimated membership of the new C.S. is about 42 000; the constituent memberships are: C.S., 17 000; R.I.C., 27 000; S.A.C., 2300; F.S., 2500; total 48 000. (About 5000 members belong to more than one body.)

The essence of the scheme was that the memberships of the C.S., R.I.C., F.S., and S.A.C shall all become Fellows of the C.S. (as modified by the scheme) under the C.S. royal charter. The C.S. in its new form will be responsible for all activities other than the strictly professional and qualifying ones of the R.I.C., which will remain separate and be the sole responsibility of the R.I.C. in its traditional role. This arrangement clearly distinguishes between learned society and professional activities, or, in other words, between the science and profession of chemistry. The R.I.C., which at that time carried out both types of activity, as did the I.P.P.S. in the case of physics, defined the position as follows:

> The organization of the science and profession of chemistry in the United Kingdom has, in common with that of other learned professions, eight basic objectives:

1. To extend and refine the fields of pure knowledge on which it is based, and to disseminate this knowledge.
2. To widen and improve understanding of the practical applications of that knowledge, and to disseminate this understanding.
3. To provide channels of communication between chemists for their mutual benefit in professional and social relations.
4. To set up and maintain adequate standards of theoretical knowledge and practical competence by establishing examinations and prescribing qualifications.
5. To ensure that suitable persons are able to receive the necessary education and training and to ensure an adequate supply of properly trained entrants to the profession.
6. To promote or safeguard the status and general interests of chemists in their relations with other professions and the general public.
7. To further — and, when necessary, protect — the legal rights, economic interests, conditions of practice or employment, and, in general, the material welfare of chemists.
8. To set standards of ethical conduct and to uphold compliance with such standards by disciplinary action.

In to-day's circumstances a ninth objective might be added, namely, to make representations to Government and give evidence to committees and commissions concerned with matters of moment to chemistry, or the profession of chemistry.

Sir David Martin (1965), the Executive Secretary of the Royal Society, has described his views of the role of the learned societies. The original role as a forum for impartial discussion between scientists and publication of new knowledge continues to be of prime importance, but a number of others have achieved importance. The central motive of all such societies, according to Martin, is to promote science for its own sake. Personally, I would prefer the Royal Society Charter motive, 'the promoting (by the authority of experiments) the sciences of the Natural things and of useful arts'. They maintain standards of integrity and excellence which have gained universal acceptance and Government recognition in the form of financial support and a dependence on them for impartial advice.

Science is international and 'as the volume of research grows, so it becomes necessary for scientists to act internationally and to meet one another more frequently'. A number of bodies have been created to this end, for example the International Council of

Scientific Unions (I.C.S.U.) and the European Physical Society. Established societies and international bodies like I.C.S.U. also assist developing countries in establishing scientific bodies of their own.

In reference to the above considerations, it will be realized that the future activities of the R.I.C. will be mainly in categories 3, 4, 5, 6, 7, and 8. The Chemical Society, on the other hand, will be concerned particularly in categories 1, 2, part of 3, and in those matters described by Martin as pertaining to learned societies.

The engineers, physicists, and chemists have taken bold steps in respect of collaboration. There is also increasing co-operation among other learned and professional bodies and some sharing of costs of common objectives, and it is certain that such palliatives will be increasingly explored. In its simplest form this co-operation has resulted in certain large societies providing secretarial and other facilities for smaller organizations which have, nevertheless, retained their own identity and independence. Examples are the facilities provided by the I.o.P. for the Association of Hospital Physicists and the Research and Development Society, and the computer facilities provided by the C.S. and the I.E.E.

14
Protective associations

THE decision of the chemical body negotiators to retain the R.I.C. charter for all the requirements of a qualifying association and a professional institute was a wise one. There are many scientists who, while enthusiastic about learned society activities, have no wish to play any part in professional activities and would regard participation in anything which smacks of trade-union activities as anathema. This, however, is not a unanimous view, especially among the younger generation of the scientific community. Scientists, like all other members of society, caught up in inflation and economic stringency, are increasingly concerned with maintaining their standards of living, as well as their professional status.

It is a surprising fact that although membership of professional bodies is not, with few exceptions, obligatory, a university degree being equally effective, the numbers in professional and qualifying associations have grown steadily. Many predicted otherwise in view of the increasing graduate population, but they have been confounded. This merits closer examination, for it indicates a strong motivation.

Prandy (1965) published the results of a study of scientists and engineers, and particularly of the conflicts between professional activities and trade unionism. The bodies he chose for his research were the Institution of Metallurgists, the Engineers' Guild, and the Association of Scientific Workers. His statistics were obtained by postal questionnaires and personal interviews. According to Prandy, scientists and engineers are middle-class, professional people, which involves them in one set of values, but the bulk of them are also employees, which involves them in a different set of values. British society is stratified by class and status, the two being closely related, but in practice inextricably confused.

Stratification by class is basically the division between those with and those without power, particularly economic power, whereas status is based on the differential distribution of honour and prestige. Status arises from the attempt of those in power to justify and legitimize their position. At one extreme there is acceptance by society of the ideology of the ruling group; at the other extreme, rejection and the challenge of power derived from class-consciousness.

Prandy believes that the factor determining acceptability or rejection is the employment situation of the group. If it shares in the exercise of authority it accepts; if its subordination is overt and emphatic, it objects or rejects. The position of the scientific community is anomalous in a number of respects. They, like engineers, regard themselves, and are accepted, as a professional class. Traditional professionalism, however, related to fee-paid practitioners. Scientific professionalism is different. There is not, except in a minority of cases, a client–practitioner relationship, as, for example, there is with lawyers or surgeons, and the scientific professional association neither confers a licence to practise, nor exercises an occupational 'closed shop'.

Since the bulk of the scientific community today are employees, their relationship to employers is a factor of increasing significance and could bring about a situation of collective conflict of interest leading to trade unionism.

In private industry, which employed nearly half of the scientists and engineers, the trend to trade unionism among scientists and engineers in 1965 was weak, but the trend was definitely greater in the case of technicians, whose participation in management was minimal. There were only two bodies of significance: the British Association of Chemists, whose membership of about 2000 (half of whom were qualified professionally) covered only some 5 per cent of the chemical industry, and the Association of Scientific Workers, with a membership of around 17 000, most of whom were not fully qualified professionals. Since Prandy's work, the Association of Scientific Workers has merged into a new body, the Association of Scientific Technical and Managerial Staffs, and it is, consequently, no longer exclusive to personnel in the scientific field.

In relation to the position of technicians in the scientific community, it is worth pointing out that in 1968 there were over 700 000 persons working as technicians or technical supporting staff (*Sta-*

tistics of Science and Technology, H.M.S.O. 1970).

We have seen that the Victorian era was characterized by the rise of the professional classes in Great Britain, which was achieved largely by the development of the professional qualifying associations. The technicians, a new aspiring middle class, who are so vital for our current industrial prosperity, are now pushing for professional status. This is a worthy aspiration which should be encouraged by the existing professional classes. Alas, in too many cases they have used their entrenched positions to deny to the new aspirants the position which they themselves achieved in the last century. Something has been done, but too little, and the alternative is to drive the technicians into trade unions.

The position in public services, including central and local government, the nationalized industries, and education, was, and is, quite different. Something like 70 to 80 per cent of scientists and engineers employed in these services are estimated to be members of a trade union, but as a class they comprise a very small percentage of the unions concerned.

Prandy's findings are borne out by other observations, and it can be concluded that the ingrained British desire for professional status is as strong as ever and that the scientific population is prepared to pay a price for it. It is also probable that the bulk of scientists and technologists still believe their professional welfare and conditions of employment can be better dealt with by a professional association, rather than, say, a trade union. Nevertheless, the climate of opinion is changing, and increasingly members are demanding more effective action on their behalf by their professional bodies.

This demand may arise from two major causes, namely, that scientists and engineers have failed to retain their salary differentials relative to other social groups and are catalysed by inflation to more militancy and, secondly, that more are coming from working-class homes with a strong tradition of trade unionism. Until recently the bulk came from the middle class which, traditionally, has had an aversion to industrial action.

The Chartered Engineering Institutions and bodies such as the R.I.C. and I.P.P.S. have for some years commissioned employment and remuneration surveys which have been most useful from the point of view of establishing reasonable remuneration levels. Recently the Ministry of Technology (as it then was) has collaborated with the C.E.I. and C.S.T.I. in surveys of professional scientists, the latest available being for 1968 (*The Survey of Professional Scientists 1968*, HMSO).

There is, however, a growing demand for more action on remuneration and conditions of employment, even for collective bargaining, and the professional institutions have been forced to give such matters close consideration. The charters of most professional bodies exclude trade-union-type activities, but the R.I.C. charter exceptionally does contain a reference which comes near to it, namely,

> That Government Departments, Local Authorities, Corporations, Public Companies, Industrial Concerns and others who use the services of persons qualified in Chemistry are recognising in increasing measure the value of the advice and information which the Institute is able to give in matters relating to the conditions of employment of such persons so that their special knowledge and experience may be used to the best advantage of the community.

The R.I.C., sensitive to the current ferment of its members, grasped the nettle and decided to try to cater for some of the rising demands. On 1 October 1970, it issued an Official Memorandum entitled 'The Future Role of the Institute as a Professional Body' and indicated its intention to strengthen its functions in respect of conditions of employment and remuneration of its members. When the R.I.C. Council issued this document they were aware that the Government intended to introduce new legislation to regulate, or set a code of behaviour for, industrial relations, and that professional scientists would be likely to be affected by such legislation. It was for this reason that the R.I.C. Memorandum contained no final proposals. The Government's White Paper (*Industrial Relations Bill—Consultative Document*) became available shortly after the R.I.C. Memorandum was issued and the Institute's Council considered that it contained nothing to alter their proposals (*Chemistry in Britain*, 1971).

The proposed Industrial Relations Bill, so far as its intentions were clear, appeared to contain aspects inimical to the interests of professional men, and the C.E.I. and C.S.T.I., along with other professional organizations, pressed the Government for amendments. A number of these were accepted. One view of the situation was expressed by the Executive Secretary of the I.P.P.S. (Cohen 1971), as follows:

> The Industrial Relations Bill is likely to prove to be the greatest stimulus to white collar unionism we have ever seen in this country and many scientists and technologists will probably find themselves involved, whether they like it, or not, in trade unionism when the Act comes into operation.

From its formation the C.S.T.I. had been preoccupied with this thorny problem of the urge towards trade unionism in the scientific population, and it set up a Working Party on Trade Unionism for Professional Scientists and Technologists. As a result of the recommendations of the Working Party, the C.S.T.I., supported by the Councils of its Constituent Bodies, decided to create 'The Association of Professional Scientists and Technologists' (A.P.S.T.) to provide a service on behalf of individual members or groups of members. A.P.S.T. was incorporated under the Companies Act early in 1972 and is registered as a trade union under the Industrial Relations Act. It is a negotiating body and protective association for professionally qualified scientists and technologists in salaried employment. A.P.S.T. will be primarily concerned with scientists and technologists in private industry, since those engaged in Government Service and Education are already covered by Professional Associations. For good, or ill, however, it appears to be a milestone in the 'unionization' of the scientific community. This is likely to intensify because, as some observers have forecast for years, the market for scientists and technologists has suddenly changed from a seller's to a buyer's market and it is difficult to find jobs. There is, in addition, growing redundancy among the middle-aged and older scientific population.

This is an extraordinary state of affairs when it seems only yesterday that Britain was making strenuous efforts to prevent and reverse the brain drain to the USA and to increase the supply of scientists and technologists by every possible means: by expansion of universities; by the creation of C.A.T.s, and, later, technological universities, and so on.

15 Salvation through science and the aftermath

THE belief in salvation through science and technology, which began during the First World War, accelerated during the Second World War and reached a peak in the early post-war years. To quote from *The New Romans* (Purdy 1968), 'The many shallow streams have widened into one great lake, the defining element of which is the belief in affluence through technology'.

The scientific method applied in the physical sciences via 'reductionism' (that is, the breaking down of matter into basic elements, molecules, atoms, and so on) had such a fantastic success that a widespread belief in an affluent society based on science and technology arose. I do not wish to decry this aspiration, which was a continuation of the Victorian dream, because without achieving at least modest affluence mankind's lot is miserable indeed.

After the war and, indeed, long before that, Britain was losing place as an industrial nation. There were mounting warnings about the technological gap with the USA, and there was even an American pamphlet entitled *Technical Stagnation in Great Britain* (Machinery and Allied Products Institute 1948). The American thesis was that the British failure was a failure of technical re-equipment, and that, having equipped its industries magnificently at the start of the modern industrial era, Britain failed to re-equip them and keep them up to date technically. The British were considered to be opposed to technical innovation or, at least, reluctant. Britain took all this seriously, perhaps too seriously, because it confused the nation, especially in relation to the persistent postwar balance of payment problems.

The belief that Britain's difficulties were primarily due to a failure of its science and technology is false, as is also the assumption that they can be put right by science and technology. Industrial success and excellence stem from achievement motivation, the spirit and mettle of a whole people, and have very many components of equal importance.

British industry itself began to have doubts about salvation through technology in the early fifties, but even as late as 1968 Government preached Mr. Wilson's red-hot technological revolution and salvation by Mr. Wedgwood Benn's Ministry of Technology. The assumption, particularly just after the war, was that if enough research and development (R and D) were carried out everything would be all right. America was a prime example of this view.

In 1964–5, the U.S.A. was devoting 3 per cent of its gross national product (GNP) to R and D. Over twenty years R and D expenditure had been increased fifteenfold, whereas GNP had only trebled. The growth phenomenon dates from the war and was prolonged after it. Before the war R and D was only 0.3 per cent of the GNP (O.E.C.D., 1968). Most American observers of these trends agreed that the R and D share of the GNP could not continue to sustain the same growth rate. Total expenditure in the U.K. on R and D has been about 2.7 per cent of the GNP in the years 1961 to 1968 (*Statistics of Science & Technology* 1970).

These figures, of course, reflect the growth of the UK scientific population from less than 500 in 1851 to around 175 000 in 1966, the growth being exponential. Similarly, the increase in the number of scientists in the U.S.A. from 46 000 in 1930 to 410 000 in 1963 (O.E.C.D. 1968, p. 167) is exponential. Such growths could not continue indefinitely and it seems as if a levelling-off is taking place, both for economic reasons and because of current public sentiment in respect of science and technology.

It is abundantly evident that science and technology, and particularly innovation derived from R and D, is being challenged and questioned. The purse-strings are being pulled tight, and value analysis and all sorts of techniques previously applied to industrial and work affairs are now being applied to R and D projects. In some extraordinary fashion the bottom has dropped out of this 'heroic materialism' (to borrow a phrase from Kenneth Clark), and it looks as if almost overnight the public lost faith in science and technology. Scientists and technologists are being made the scapegoats for the social and industrial ills which beset our society.

The accusations are that science and technology, and particularly R and D, are responsible for producing our present industrially based society and the 'soulless' megamachines of large-scale industry. In particular, technological innovation in industry has been pursued regardless of the social and amenity consequences and without respect to the conservation and maintenance of a tolerable environment. They are also blamed for providing the horrific sinews of war culminating in the atomic bomb. It is high time that the science and technology community stopped feeling so guilty and said a word or two in their own defence.

The year 1851 is regarded by many as the peak of Great Britain's industrial achievement: the year of the Great International Exhibition at the Crystal Palace, which was the envy of the whole world. Britain may have been at the peak of its industrial achievement, but there were black ash wastes in Lancashire, Newcastle, and Glasgow; there were furnace slag and copper furnace waste dumps in South Wales; there were pit-heaps and tips throughout the country, and the Thames presented probably the finest stretch of effluent in the world. There were pea-soup fogs and inversion layers; there were magnificent plumes of chimney effluents, and where there was brass, there was muck in profusion. That was the picture in 1851 and, as we saw earlier, the 1851 Census lists 442 'scientific persons' in a total population of about 18 million. Less than 500 persons, some would have us believe, created all that pollution. The facts clearly establish that those responsible for fashioning our present industrial society were certainly not scientists. The 'know-how' which developed industry in Britain was a complex mixture of arts, crafts, and technology, or, borrowing Lewis Mumford's nomenclature, technics, going back to ancient times. It was only in fairly recent times informed by science.

So much for Victorian industrial Britain, which we have inherited and which is with us, whether we like it or not. The fact is that men have been polluters since the beginning of time. The noble savage, who conserved his environment, is a myth. Nomads polluted and moved on and, in due course, nature recovered the situation. 'Primitive villages', to quote a description from a novel of Algeria in the early days of the Knights Templars, 'melting one into the other, left an impression only of filth and smell, and a scattering of chickens and goats, and naked children, smeared grey with mud' (Dunnett 1971, p. 104).

There is another side to this depressing picture, for the technics of man in agriculture, horticulture, and architecture, in other words the ancient sciences of mankind, have transformed many natural habitats into pleasant places. The countryside in Britain, which we are so desirous of conserving, is, for the greater part, man-made. Anyone who has a garden knows what happens when nature is left to herself and rank and rampant growth prevails. Nor is it only men who create deserts — nature itself has produced some prime examples.

While it is reasonable to repudiate the scientific community's responsibility for the gross errors of the industrial revolution, they cannot escape the consequences of the results of the scientific discoveries which have arisen in recent years. I can give one clinching example of this. Physicists, in particular, long believed that they had no responsibility for the practical consequences of their discoveries. The atom bomb changed all that as the route from fundamental science to universal destruction became all too clear. Responsibility for the social consequences of R and D is now shared by scientists with others. Science is no longer a thing apart; it is part and parcel of our social order and its objects are multivalent.

16

Towards a better future

I believe we are at the beginning of changes as important as those initiated in the Classical Era of Greece, or the New Learning of the Elizabethan period. These are concerned with our attitudes and motivation in respect of knowledge and its application. There is a growing belief that people matter more than things.

The growing awareness of the public in respect of deterioration of our environment and of the prodigal waste of our natural resources has led to an anguished questioning of the technologically based industrial society. Man's fear of machines, which in the last century produced Luddites, has turned to a fear of science and technology, and the modern Luddites wish to break the mega-machines of industry, Government, and establishment. People, especially the young, have become disenchanted with science and technology and are turning away from it to seek via anarchy and violence an alternative culture, rather than face the world as it is.

The situation unfortunately cannot be remedied by opting out: by emotional retreat or violence. Sooner or later we shall have to accept that the world is as it is, and do what we can about it. It is as we have made it, all of us, and it behoves us to have a hard cold look at ourselves — not for self-pity, or to blame others — but to pull ourselves up by our own bootstraps.

The public are not much longer going to tolerate industrial and economic innovation without regard to its social and ecological consequences. They will demand that improvements in the material standard of living be coupled with quality achieved in an improving ecological environment and with organizations that are more socially acceptable and tolerable. These are the inevitable demands of an opulent society. They will necessitate a profound modification of Government and industrial motivation and the deployment of more resources to study the likely consequences and side-effects of new projects, products and processes. This, in turn, will require great efforts of an integrated type embracing many disciplines.

First of all we must accept, and this is extremely difficult, that salvation by material affluence only is not the answer. It has been tried and failed, though obviously it is a necessary part of the answer. People on the bread-line with empty bellies are not fulfilled, or happy; they are either apathetic or bitterly resentful.

We appreciate now that for the good life, somehow or other, affluence must be coupled with quality. The pursuit of quality, or excellence, as well as material abundance, presents new opportunities for the scientific and technical community, though not in isolation.

I do not believe that affluence and quality will solve our problems though they are a vital part of the solution. People have deep emotional needs, and social and industrial systems based solely on affluence and quality will still be unable to sustain happy people adjusted to their environment. Letting men and women live their lives in the most fulfilling ways is our objective, but we do not know how to develop what might be called acceptable, or perhaps less ambitiously tolerable, systems. We shall have to try and find out.

On a broad canvas the Bishop of Durham (Ramsey 1971), in an address entitled 'The influence of technology on the social structure', gave a masterly exposition of the issues involved. He said:

> We are witnessing in our day a fundamental change in the attitude of science to morality, and a new awareness of the significance of technology and of its important possibilities for society. There is before us for that very reason the possibility of a new culture, scientific and moral, technological and Christian, a possibility for which we have waited for two, if not three, centuries, a possibility denied us earlier by an unholy non-aggression pact between science on the one hand, and morality and theology on the other. Alongside this possibility the problems of to-day, deep and far-reaching though they be, can be understood as the birth pangs of a new era; they are certainly matters of life and death for man and society.

What is the position of the scientific community in the new culture which is developing? Man's concern about his environment has been growing over many years, but during the Second World War and the years directly afterwards there was little chance to do much about it. The catalyst to positive action was probably supplied by Rachel Carson, herself a scientist, who from 1958 had been studying the effects of man-made poisons, particularly insecticides, on the environment and the living community. Her book *Silent Spring* (Carson 1962) came as a shock, not only to the general public, but to the scientific community, many of whom thought her views exaggerated, but nevertheless commanding attention. Further rude awakening of public opinion occurred after the *Torrey Canyon* disaster in 1967 and the poisoning of the Rhine in 1969.

In May 1963, which included National Nature Week in Britain, the Wild Life Exhibition created much concern and H.R.H. the Duke of Edinburgh, himself a great befriender of causes, founded the movement called 'The Countryside in 1970'. The object was to promote what might be called a meaningful conservation lobby for appropriate action.

Three conferences have been held: in 1963, on three broad issues: conservation, organization, and research; in 1965 on specific proposals for future development; in 1970 a review of the work on the past seven years and of proposals for action under six headings as follows: (1) agriculture and forestry; (2) urbanization and the countryside, (3) industry and the countryside, (4) social issues, (5) leisure, (6) responsibilities for the environment: (A) advisory services, (B) Government and voluntary bodies. The Standing Committee also participated in the European Conservation Year and other international activities. The review of the third conference concludes:

> Overall, one feature emerges, that there can be no obvious and simple solutions to the complex and difficult problems of environmental management. It is the realisation of the inter-relationship of so many of these issues and that continuing team work and mutual understanding are essential, which makes for progress.
>
> And this must be on a global scale, for our world environment is a unity. Almost any action in one part affects all of us to some extent, so that we have to examine our problems in an international context.

The R.S.A. and the Nature Conservancy had jointly provided the secretariat and administration for 'The Countryside in 1970'. At the conclusion of the third conference, the chairman put forward a resolution:

That this Conference notes with satisfaction the increased interest shown in the environment by public opinion, recognises the critical situations and problems which yet remain to be resolved, and therefore welcomes the inclusion by two successive Prime Ministers of a Minister in their Cabinets specially responsible for the environment.

It supports the activities developing from European Conservation Year, thanks the Royal Society of Arts for accepting responsibility for convening from time to time meetings to continue this work done by the 1970 Movement during the last seven years, and welcomes the continuing interest in environmental issues of His Royal Highness, The Duke of Edinburgh, in his capacity as President of the Royal Society of Arts.

In this way the R.S.A. has undertaken a continuing responsibility for conservation matters. In a manner of speaking, it has gone full circle because it actually was the first amenity society. In 1763 it offered prizes for 'devices to swallow smoke from factory chimneys in London'. The project was not successful because it was too far in advance of the times.

The Government has reacted to public anxiety and created a Department of the Environment headed by a senior Secretary of State.

Conservation problems are international and this is increasingly recognized. European Conservation Year (1970) brought European thinking together in conferences. The Committee on Environmental Conservation (Co.En.Co.), the Council of Environmental Education, and the Environmental Group of the National Union of Students are all active in the international field. In December 1970 Lujo Tončić-Sorinj, Secretary-General of the Council of Europe, issued a letter to Europeans from the Council of Europe detailing the problems in simple terms and asking for co-operation and action on a number of practical suggestions. On the wider international scene the United Nations Economic Commission for Europe has a conservation programme on man and his biosphere, and the United Nations held a World Conference on the human environment in Stockholm in June 1972.

Even in Britain alone it will be extremely difficult to obtain agreement as to what should be done, owing to the variety of interests involved, many conflicting. There were, for example, no less than 335 organizations co-operating in 'The Countryside in 1970'.

There has always been, and always will be, a necessity to balance the needs and claims of different sections. How can we strike a fair balance between the material standard of living and the immaterial standard of living, or, as Peter Scott put it, the quality of life, rather than affluence? To a great extent this must be a matter for Government, but unless people feel enough, and care enough, little will be done. This struggle for a balance is not just a struggle between the economists and the ecologists. It is of vital interest to everyone, and the public should be adequately and correctly informed on such serious matters so that it can understand the choices which are open. Some say, of course, that the whole business is being blown up far too much, but the Secretary-General of the Council of Europe rejects the criticism that many people are over-dramatizing the effects of man's interference with his environment. What is being done, he claims, is of such a nature, and on such a scale, that natural forces cannot rectify, or compensate, as they have been able to do when changes were relatively slow.

Industry, which is often considered the villain of the piece, is a human activity — and contains a large cross-section of society which is largely responsible for our affluence. We must face the fact that the approach to these problems is primarily one of self-interest and that issues are often fought to the last ditch — ditches often so shallow that people should never have got into them, let alone die in them. The scientific community, and its learned societies and institutions, has a vital and indispensable part to play in this great enterprise which involves the conservation and repair of our environment.

New horizons 17

INDIRECTLY science and technology have produced our current ills. Circumstances are now generating the emotion needed to produce a public urge to get something done, but only the scientific community has the specialist knowledge which alone can provide the solutions in terms of hard technology and economics.

Although we live in an age of intense specialization, and necessarily so, all our current big problems are generalist problems requiring concerted action by people of wisdom in multiple disciplines. No specialist body, whether it be a learned scientific society or an engineering institute, can act alone to solve our current problems. A multi-disciplinary approach is needed and difficult compromises have to be made. Many of the problems are international in character and will need world-wide collaboration.

In industry, scientists will increasingly be concerned with controlling and reducing effluents and in designing new plants, engines, and systems which perform at a greatly increased standard of cleanliness. Such quality R and D should appeal particularly to the young, and may stimulate some of them to undertake the hard discipline of learning science so as to fit themselves to tackle the problems of improving our environment, controlling noxious effluents, and optimizing the benefits of science for the good of all. Such work could well supplement the pursuit of knowledge for its own sake and supply a new and powerful emotional motivation to replace the current disenchantment with science and technology.

The multi-disciplinary approach will require a new outlook, not only for scientists, both in industry and elsewhere, but in the learned societies and institutions. Learned societies were originally generalist societies, and it is probable that the day of the generalist society is coming again.

It is doubtful whether the environmental problems can be solved without population reduction or, at least, by diminution in the rate of increase. This, of course, has emotional and religious overtones which are difficult to deal with, but the problem is outstandingly one which requires concerted action. Pharmaceutical and social research are two elements involved in a multiplicity of requirements to effect this end.

Further, it is almost certain that conservation problems will remain obdurate unless individual demands on world resources are reduced. This goes against all current demands and theories in a consumer society of contrived needs. It would involve a return to the old philosophy in which things were made to last and quality was more important than quantity. We have been 'conned' for so long into changing our refrigerators or cars every time there is a minor change in fashion that it will be hard to break the habit. The economic consequences of contained production are profound, as also are the industrial problems involved, but there is no theoretical reason why systems based on quality production, rather than mass production, should not be viable. Sooner or later a lot of research will have to be done in this field. Criteria for survival, therefore, may well be fewer people with less individual demand on limited exhaustible resources.

The specific contributions which the scientific community can make towards the solution of the current problems we have considered are:

(1) To continue their efforts in producing material affluence by innovation.

(2) To guide innovation and R and D towards improving the quality of life as well as affluence.

(3) To study the reaction of systems (including social and industrial systems) when confronted by innovation, with the object of developing dynamic systems which are tolerable and eventually fulfilling.

Category (1) is the main function which scientists already fulfil. So far, however, it has been done in a fairly narrow confinement of physical science, with little consideration until recently of (2) and (3). Categories (1) and (2) are more or less self-explanatory and there is no need to go into great detail, but category (3), the reactions of systems, is a newer conception which needs some further consideration.

Systems may be natural or man-made, but all are characterized by having some function or purpose imposed upon or inherent in them. A tree or a man is a system designed to perform a function or purpose which cannot be determined by an examination of the component parts. Every forest, stream, and country is a finely balanced ecological system.

Animals and insects, such as bees and ants, have produced communication and social systems of considerable merit and complexity, and in the case of the lower orders, as far as we know, without any conscious thought. Man-made social systems, including families, communities, institutions, and industrial concerns, are the outcome of both conscious and unconscious purposes, though in what proportions it would be difficult to say. Such systems react when exposed to, or confronted by, innovation, and the reaction is to a large degree illogical and emotional, arising from an instinctive wish to maintain the status quo.

Scientific and technological discoveries and inventions have no effect until they are introduced into systems. For example, it is only when the steam engine is used in some system, e.g. a railway network, that a direct effect ensues. The innovation initiated by the railway system confronts all man's social systems, and they react to it one way or another.

Pursuing the subject into the industrial sphere, a new expensive machine or process put into a factory will be useless if the work-force refuse to operate it, or its advantages will be seriously frustrated if they work it ineffectively. In a word, technological innovation can be effective only in the context of a favourable reaction by the system (in this case a work-force) which is confronted by it. As a consequence R and D into the confrontation reactions of the receiving system is just as necessary as, sometimes more necessary than, R and D on the innovation itself.

The success of the scientific method led people to adopt analogous methods (albeit sometimes superficial ones) for other studies, one of which was work and methods study. Now there is no doubt that systematic study of operations and methods in manual work was much needed and inevitable. The method of implementing measurements by incentive bonuses paid to each worker on his individual performance is, however, quite another matter, and however logical this might appear to be it is fraught with desperate consequences for the future. It can be argued that the long-term effects of individual bonus systems on close-knit work-force social systems could have been foreseen and that short-term gains were gained for the price of long-term disruptions, but the fact is that few technical people at that time knew that in such confrontations emotion would always triumph over logic. We have learned this the hard way. A recent example is the flat refusal of some docks to accept containerization. Similarly, the deep-seated problems of the railways, the coal-mines, or the shipyards cannot, much as we would like it, be solved simply by the application of logical and scientific plans of innovation. R and D, like politics, will have to be concerned more and more with the art of the possible, as far as innovation is concerned.

It would appear essential that an experimental approach should be adopted and individual experiments be carried out to find out what sort of systems are tolerable in industry.

Facilities for carrying out experiments could certainly be created if natural and social scientists and others co-operated. The difficulties of such an approach must not be under-estimated, for it would go much against the grain of established positions, prescriptive dogmas (such as collective bargaining), and centralized bureaucracy, to mention but a few. It has, however, in my view, the germs of a new 'philosophy', in which the experimental approach of the new learning would be turned not only to 'the patient investigation of nature by observation and experiment', but to letting men and women live their lives in the most fulfilling ways.

If the scientific community were relieved, in the words of the Bishop of Durham quoted above, of the 'unholy non-aggression pact between science, on the one hand, and morality and theology, on the other', and of the equally unholy pact to keep out of politics, it can play a full part in the new philosophy. Indeed, a clue to a practical philosophy for the scientific community is already to be found in the medical community, whose avowed object is healing and preventing ill-health, whether by medicine, surgery, or (more recently) psychology. The aim is to promote the well-being of man physically and mentally. It is true that much research work is carried out in biochemistry, biophysics, pharmaceuticals, and so on to advance knowledge, but the prime purpose of these investigations is towards producing cures, or to prevent illness. It is not suggested that research workers in the medical field are not interested in advancing knowledge *per se,* but their basic motivation is practical application.

The motivation of the scientific community in the future will be increasingly towards (1) the pursuit of knowledge and (2) its application for the material and moral benefit of man and his environment. Certainly the motivation of a privileged minority will continue to be (1), but for the majority it must be (2), just as it is in the medical world.

In respect of politics, the scientific community must rethink its position.

There has been in Britain an almost complete exclusion of scientists and technologists from the processes of Government, from the real corridors of power. This is too precise a situation to be accidental, and in the view of many people it is a deplorable state of affairs in a technologically based society. Sampson (1962) remarks that 'There is no greater, or more difficult, challenge to modern Governments than the integration and organization of scientists into the fabric of administration. Britain has excelled in producing these awkward, independent men; but has kept them insistently apart from the traditional institutions of her Government.'

Historically, positions in Government have been acquired by people who had power themselves, either inherited, or acquired. The governing class does not readily accept dilution of power, and new entrants have to force their way in by pressure groups. They are successful only to the extent that they have power themselves or are backed by power.

Those participating in Government until recently were the old-fashioned ruling classes, the professions, and the independent employer, or owner of industry, who acquired power during the industrial revolution. The worker and his trade union representatives were excluded until they acquired power and influence from the organization of workmen into power groups and could command votes from universal suffrage.

The traditional ruling classes had little respect for science, technology, and industry; and what we call society — an amorphous body which cannot be defined, but which we all know exists — shared these views. The new categories that entered Parliament and acquired power in Government had mounted great efforts to do so. No one put out red carpets for them; they knew what they wanted and they fought to get it. It is equally clear that the scientists and technologists will not be accepted, except in subsidiary roles, unless they themselves will otherwise and fight for it.

References

ALLAN, D. G. C. (1968). *William Shipley, founder of the Royal Society of Arts*. Hutchinson, London.

ANON. (1964). *Scientific and learned societies of Great Britain*. Allen and Unwin, London.

ASTBURY, N. F. (1968). The research associations: fifty years of applied science. *Bull. Inst. Phys., Lond.* **19**, 212.

CARSON, R. (1963). *Silent spring*. Hamish Hamilton, London.

CHEMICAL SOCIETY (1896). *The Jubilee of the Chemical Society of London 1891*. Chemical Society, London.

CHIRNSIDE, R. C. (1971). The analytical revolution — its impact on academic and industrial research. (Eighth Dunn memorial lecture.) *Chemy. Ind.* **33**, 893.

COHEN, H. (1971). *Phys. Bull.*, **22**, 319.

COTGROVE, S. and BOX, S. (1970). *Science, industry and society: studies in the sociology of science*. Allen and Unwin, London.

Department of Education and Science, Ministry of Technology (1970). *Statistics of science and technology*. Her Majesty's Stationery Office, London.

DUNNETT, D. (1971). *Pawn in frankincense*. Cassell, London.

HODGES, H. (1970). *Technology in the ancient world*. Penguin, London.

KITTO, H. D. F. (1969). *The Greeks*. Penguin Books, London.

Machinery and Allied Products Institute (1948). *Technical stagnation in Great Britain*. Machinery and Allied Products Institute, U. S. A.

MARTIN, D. (1965). The role of the learned societies. *Discovery, Lond.* **26** (9), 16.

MILLERSON, G. (1964). *The qualifying associations*. Routledge and Kegan Paul, London.

MUMFORD, L. (1967). *The myth of the machine*. Secker and Warburg, London.

MYERS, A. R. (1969). *England in the Middle Ages*. Penguin Books, London.

Organisation for Economic Co-operation and Development (1968). *Reviews of national science policies. United States.* Directorate for Scientific Affairs, O.E.C.D., Paris.

PACKMAN, R. (1965). *A guide to the business world.* Longmans, London.

PILCHER, R. D. (1914). *History of the Institute of Chemistry of Great Britain and Ireland 1877-1914.* Institute of Chemistry of Great Britain and Ireland, London.

PORTER, G. (1971). The amalgamated societies. *Nature, Lond.* Chemical Society supplement (31 Dec. 1971).

PRANDY, K. (1965). *Professional employees — a study of scientists and engineers.* Faber & Faber, London.

PURDY, A. W. (1968). *The new Romans.* M. G. Hurtig, Edmonton, Canadá.

RAMSEY, I. T. (1971). *The influence of technology on social structure. Jl R. Soc. Arts,* **119**, 602.

READER, W. J. (1966). *Professional men. The rise of the professional classes in nineteenth-century England.* Weidenfeld and Nicolson, London.

RILEY, J. F. (1954). *The hammer and the anvil: background to Michael Faraday.* Dalesman, Lancaster.

ROYAL SOCIETY (1912). *Record of the Royal Society, 3rd edition.* Royal Society, London.

SAMPSON, A. (1965). *Anatomy of Britain today.* Hodder and Stoughton, London.

SHILS, E. (1968). The profession of science. *Advmt Sci., Lond.* **24**, 893.

SMITH, R. (1971). *Chemy Brit.,* **7** (5), 185.

TAYLOR, J. (1961). The modern chemical industry. *Jl R. Soc. Arts.* **109**, 356.

TAYLOR, J. (1967). 'Mergers' and 'marriages'. *Chemy Ind.* Feb. 11, 213.

THRING, M. W. (1965). The nature of pure and applied physics. *Bull. Inst. Phys. Lond.* **16**, 305.

TRUEMAN WOOD, H. (1913). *A history of the Royal Society of Arts.* John Murray, London.

WALKER, E. A. (1969). The work of the National Academy of Engineering, U. S. A. *Jl R. Soc. Arts.* **117**, 385.

WECK, R. (1967). Co-operative research in competitive industry. *Advmt Sci. Lond.* **24**, 129.

WILLIAMS, L., JONES, E., and TAYLOR, J. (1969). Amalgamation. *Chemy Brit.* **5** (10), 436.

WOOD, H. P. (1968). From Robbins to 1970. *Advmt Sci., Lond.* **23**, 65.

Index